Hanno Homann

SAR Prediction and SAR Management for Parallel Transmit MRI

Vol. 16
Karlsruhe Transactions on Biomedical Engineering

Editor:
Karlsruhe Institute of Technology
Institute of Biomedical Engineering

Eine Übersicht über alle bisher in dieser Schriftenreihe erschienenen Bände finden Sie am Ende des Buchs.

SAR Prediction and SAR Management for Parallel Transmit MRI

by
Hanno Homann

Dissertation, Karlsruher Institut für Technologie
Fakultät für Elektrotechnik und Informationstechnik, 2011

Impressum

Karlsruher Institut für Technologie (KIT)
KIT Scientific Publishing
Straße am Forum 2
D-76131 Karlsruhe
www.ksp.kit.edu

KIT – Universität des Landes Baden-Württemberg und nationales
Forschungszentrum in der Helmholtz-Gemeinschaft

KIT Scientific Publishing 2012
Print on Demand

ISSN: 1864-5933
ISBN: 978-3-86644-800-1

SAR prediction and SAR management for parallel transmit MRI

Hanno Homann

INSTITUTE OF BIOMEDICAL ENGINEERING
KARLSRUHE INSTITUTE OF TECHNOLOGY (KIT)

in collaboration with

PHILIPS TECHNOLOGIE GMBH
INNOVATIVE TECHNOLOGIES
RESEARCH LABORATORIES HAMBURG

SAR prediction and SAR management for parallel transmit MRI

Zur Erlangung des akademischen Grades eines

Doktor-Ingenieurs

von der Fakultät für
Elektrotechnik und Informationstechnik
des Karlsruher Instituts für Technologie (KIT)
genehmigte

Dissertation

von
Dipl.-Ing. Hanno Homann
geboren in Hannover

Tag der mündlichen Prüfung: 29.11.2011
Hauptreferent: Prof. Dr. Olaf Dössel
Koreferent: Prof. Dr. Klaas Prüssmann

Abstract

Parallel transmission, i.e. the use of multiple independent RF transmit coils, enables a relatively direct control of the RF field and is considered as a key-technology to recover image homogeneity in high-field MRI ($\geq 3\,\mathrm{T}$). The use of parallel transmission has, however, also caused concerns about the specific absorption rate (SAR). Depending on the multi-channel drive, the SAR can significantly vary and the location of SAR hotspots within the patient body can change. Conventional safety assessment schemes for single-channel transmission fail to address this situation, hence more sophisticated SAR prediction concepts are required.

The present thesis aims to provide such methods for comprehensive safety assessment in parallel transmit MRI, based on finite-difference time-domain (FDTD) simulations and on measurements. This work first focuses on modeling of a multi-channel RF coil and cross-calibration to the real physical RF system. Then, a novel approach to generate human body models from an MR pre-scan is proposed, based on water-fat separation. A first *in vivo* validation of human body models was performed by means of B_1 mapping.

Parallel transmit MRI can furthermore be applied for SAR reduction in addition to improving RF excitation uniformity. This is referred to as "SAR management" in this work. It could be shown for a broad variety of human body models that an improved uniformity is related to reduced SAR. To get a better handle on the SAR, a novel optimization approach with local SAR constraints is developed. This presents a robust and fast approach to achieve optimal uniformity while maintaining local SAR limits. Finally, the potential of SAR reduction by adapting the channel weights depending on the location in the sampling k-space is explored. This approach takes advantage of the fact that most signal information is contained in the central k-space. A notable additional SAR reduction with marginal compromises on image quality could be achieved.

As the SAR is a major limiting factor in most high-field MRI protocols, the proposed approaches to SAR management can be directly applied to improve image quality or to reduce examination times.

Zusammenfassung

Parallele Transmission, d.h. die Verwendung von mehreren unabhängigen RF-Sendespulen, ermöglicht eine relativ direkte Steuerung des RF-Feldes und wird daher als Schlüsseltechnologie für die Hochfeld-MRT ($\geq 3\,\mathrm{T}$) betrachtet. Gleichzeitig verursacht die Parallele Transmission aber auch Bedenken bezüglich der Spezifischen Absorptionsrate (SAR). Abhängig von der Ansteuerung der verschiedenen Kanäle können sich signifikante Veränderungen des SAR ergeben. Dies wird durch herkömmliche Methoden zur SAR-Abschätzung nicht berücksichtigt. Daher werden neue, fortgeschrittene Konzepte benötigt.

In dieser Arbeit werden solche Methoden vorgestellt, basierend auf Simulationen sowie Messungen am MR-System. Der Schwerpunkt liegt dabei zunächst auf der Modellierung einer Mehrkanal-RF-Spule und der Kalibrierung der Simulation in Bezug auf das physikalische MR-System. Anschließend wird ein neuartiger Ansatz zur Generierung von Patientenmodellen anhand von Wasser-Fett-separierten MRT-Daten vorgestellt. Damit konnten Simulationen des RF-Feldes erstmals im menschlichen Körper validiert werden.

Weiterhin ermöglicht die Parallele Transmission eine Reduktion des SAR bei gleichzeitiger Verbesserung der Anregungshomogenität. Dies wird hier als „SAR-Management" bezeichnet. Es konnte an einer Vielzahl von Körpermodellen gezeigt werden, dass eine verbessere Homogenität mit einer Verringerung des SAR einhergeht. Um das SAR direkter beeinflussen zu können, wird ein neuer Ansatz zur SAR-beschränkten Optimierung entwickelt. Schließlich werden Möglichkeiten zur SAR-Reduktion untersucht, die sich aus einer adaptiven Wahl der Kanalgewichte im k-Raum ergeben. Dabei wird die Tatsache ausgenutzt, dass der Informationsgehalt des MR-Signals im Zentrum des k-Raums maximal ist. Mit diesem Ansatz konnte eine zusätzliche SAR-Reduktion erreicht werden.

Da das SAR bei vielen Protokollen der Hochfeld-MRT einen limitierenden Faktor darstellt, können die hier vorgestellten Ansätze zum SAR-Management relativ direkt in eine Verbesserung der Bildqualität oder Verringerung der Untersuchungsdauer umgesetzt werden, zum Nutzen der ärztlichen Diagnostik und zum Wohl des Patienten.

Acknowledgments

I feel deeply thankful to my supervisors Peter Börnert, Ingmar Gräßlin, and Olaf Dössel. Without your knowledge, ideas, and continuous support this thesis could not have been written.

This research work was carried out in close collaboration with Philips Research Hamburg, where I spent most of the past three years. I would like to thank the head of the department of Tomographic Imaging Systems, Dye Jensen, for giving me the opportunity to pursue this work and supporting me wherever needed. I also want to express my gratitude to all the people working in the lab, in particular Peter Vernickel, Ulrich Katscher, Kay Nehrke, Christian Findeklee, and Holger Eggers, for creating such a friendly and inspiring atmosphere.

The past three years would also not have been the same without my fellow PhD students Tobias Voigt, Eberhard Hansis, Mariya Doneva, Alfonso Isola, and Nadine Gdaniec. Thanks for the good time and all the coffees we shared.

Finally, I want to thank Stefanie for reminding me that private life is more important than work.

Hanno Homann
Hamburg, September 2011

Contents

Abbreviations

ACQ	acquisition
AFI	actual flip-angle imaging
CSF	cerebrospinal fluid
CT	X-ray computed tomography
CV	coefficient-of-variation
EM	electro-magnetic
FFE	fast field echo MR sequence
FDTD	finite-difference time-domain
FOV	field-of-view
GPU	graphics processing unit
MBC	multi-transmit body coil
PUC	pick-up coil
RF	radio frequency
rms	root-mean-square value
ROI	region of interest
Rx	receive
T1	spin-lattice relaxation time
T2	spin-spin relaxation time
TR	repetition time of an MR sequence
TE	echo time of an MR sequence
TEM	transversal electro-magnetic
Tx	transmit
TSE	turbo spin echo MR sequence

Physical Constants

Symbol	Value and unit	Description
γ_H	$2\pi \cdot 42.576\,\mathrm{MHz/T}$	gyromagnetic ratio of ^{1}H hydrogen
ε_0	$8,854 \cdot 10^{-12}\,\mathrm{As/Vm}$	permittivity of free space
μ_0	$4\pi \cdot 10^{-7}\,\mathrm{Vs/Am}$	permeability of free space

Formula Symbols

Symbol	Unit	Description
$a(t)$	–	RF waveform
a_{rms}	–	root-mean-square value of $a(t)$
B_0	T	main magnetic field
B_1^+	μT	transmit-active magnetic field
B_1^-	μT	non-transmit-active magnetic field
c_H	J/K/kg	specific heat capacity
E	V/m	electric field
f_L	1/s	Larmor frequency
I		element current
i	–	imaginary unit
k	rad/m	angular wavenumber
N	–	number of RF Tx elements
N_T	–	number of sampling points in time
$\mathbf{Q}$	W/kg	Q-matrix
$\mathbf{S}_{B1}$	μT	magnetic Tx field sensitivity
$\mathbf{S}_E$	V/m	electric Tx field sensitivity
SAR	W/kg	specific absorption rate
SNR	dB	signal-to-noise ratio
V	V	voltage
$\mathbf{w}$	–	vector of channel weights ("shim setting")
$\mathbf{w}(t)$	–	vector of RF waveforms
$\mathbf{A}_{\mathrm{sys}}$	–	system matrix of the transmit chain
α	°	flip-angle
ε_r	–	relative electrical permittivity
λ	m	wavelength
ρ	kg/m^3	mass density
ρ_{el}	C/m^3	electrical charge density
σ	S/m	electrical conductivity
ω	rad/s	angular frequency

Chapter 1

Introduction

High-field MRI ($\geq 3\,$T) is gaining clinical importance due to a high signal-to-noise ratio (SNR), facilitating improved image quality and a reduced total examination time. However, at high field strengths, the wavelength of the radio frequency (RF) field ($\lambda \leq 30\,$cm) is in the range of the dimensions of the human body, leading to spatially inhomogeneous excitation profiles. This can result in local shading and undesired contrast variations in MR images.

Parallel transmission, i.e. the use of multiple independent RF transmit coils, enables a relatively direct control of the RF field and is considered as a key-technology to recover image homogeneity in high-field MRI. RF shimming is a concept for optimization of the transmit field homogeneity, using the same RF waveform but different amplitudes and phases for the different transmit channels [1, 2]. This allows recovering uniform image intensity and contrast in high-field MRI, particularly for body and breast imaging at $3\,$T and for neurological imaging at $7\,$T and above. In the more general concept of transmit sensitivity encoding (Transmit SENSE), the RF waveforms of each transmit (Tx) channel are tailored for accelerated spatially or spectrally selective excitation [3–5]. This has multiple interesting applications such as zoom imaging, arterial spin labeling, or localized spectroscopy.

The use of parallel transmission has, however, also caused concerns about RF power, which is mostly dissipated as heat in the patient body. As local temperature changes are difficult to predict, the specific absorption rate (SAR) was defined as the dissipated power per unit mass to describe RF exposure.

As the local SAR inside the patient body is difficult to measure, numerical simulations need to be performed to assess patient safety [6]. However, numerical SAR values can significantly vary, depending on the patient's anatomy. Unfortunately, only a limited number of generic body models is available for SAR simulations and the required model complexity is little understood. A broad range of human body models is needed to allow a comprehensive SAR prediction.

Depending on the multi-channel drive, the SAR can significantly vary and the location of SAR hotspots within the patient body can change. Conventional safety assessment schemes for single-channel transmission fail to address this situation, hence more sophisticated safety concepts are required. Furthermore, the potential to control the RF field raises the question how a multi-channel RF system can be employed to reduce local heating.

The present thesis aims to provide methods to address these challenging demands. In chapter 2, the state of the art in parallel transmission and SAR calculations is reviewed. Chapter 3 describes the modeling of a multi-channel RF body coil and a cross-calibration to the physical RF system. Furthermore, the coil model is validated by measurements of the magnetic and electric fields as well as by temperature measurements. Chapter 4 addresses the current lack of suitable human body models and proposes a novel approach to generate reasonably accurate body models from an MR pre-scan based on water-fat separation. A first *in vivo* validation of human body models was performed by means of B_1 mapping. Based on these models of the RF coil and the patient, a novel approach to SAR management for RF shimming is presented in Chapter 5. In Chapter 6, a new method to reduce SAR in RF shimming by adapting the channel weights depending on the location in the sampling k-space is proposed. This work concludes with a summary in Chapter 7.

Chapter 2

Parallel transmit MRI and SAR

2.1 RF transmission in MRI

2.1.1 Foundations of RF transmission and reception

Magnetic resonance imaging (MRI) is based on a quantum mechanical property of atomic nuclei, known as nuclear spin. Clinical MRI uses primarily hydrogen ^{1}H nuclei due to their high abundance in human tissues. When a nuclear spin system is exposed to an external magnetic field $\mathbf{B}$, its net magnetization vector $\mathbf{M}$ experiences an aligning force and precesses according to the Bloch equation [7]:

$$\frac{\mathrm{d}\mathbf{M}}{\mathrm{d}t} = \gamma \cdot \mathbf{M} \times \mathbf{B} \tag{2.1}$$

where γ is known as the gyromagnetic ratio. If the magnetic field is oriented anti-parallel to the z-axis, $\mathbf{B} = (0, 0, -B_0)$, the magnetization describes a right-hand screw rotation about the z-axis at a rate known as the Larmor frequency:

$$f_L = \frac{\gamma}{2\pi} B_0 \tag{2.2}$$

For hydrogen, the Larmor frequencies are $f_L = 63.9\,\mathrm{MHz}$ and $f_L = 127.7\,\mathrm{MHz}$ at $B_0 = 1.5\,\mathrm{T}$ and $B_0 = 3\,\mathrm{T}$, respectively. When an RF pulse at this frequency is applied to the spin system at thermal equilibrium, the magnetization is rotated by the flip angle α:

$$\alpha(\mathbf{r}) = \gamma \cdot \int_0^T B_1^+(\mathbf{r}, t)\,\mathrm{d}t \tag{2.3}$$

Only the circularly-polarized magnetic field component B_1^+, which rotates in the same direction as the nuclear precession, contributes to the spin excitation [7]. For efficient power transfer, Tx coils are typically designed as resonant circuits in a so-called "birdcage" geometry [8] and driven at quadrature excitation to produce a predominantly circularly-polarized magnetic field. This rotating field can be calculated from the magnetic flux density B_{1x} and B_{1y} (in the laboratory frame), generated by the currents in the resonant RF coil as (see Appendix A.2):

$$B_1^+(\mathbf{r}) = \frac{1}{2}\big(B_{1x}(\mathbf{r}) + iB_{1y}(\mathbf{r})\big) \qquad (2.4)$$

The corresponding field component rotating in opposite direction B_1^- can be calculated as:

$$B_1^-(\mathbf{r}) = \frac{1}{2}\big(B_{1x}(\mathbf{r}) - iB_{1y}(\mathbf{r})\big) \qquad (2.5)$$

For small flip angles and neglecting relaxation effects, the final transverse magnetization $M_{xy}(\mathbf{r})$ at time T generated by an on-resonance RF pulse applied in the presence of a gradient field can be found by integrating the Bloch equations as [9]:

$$M_{xy}(\mathbf{r}) = i\gamma M_0 \cdot \int_0^T B_1^+(\mathbf{r}, t) \cdot e^{i\mathbf{r}\cdot\mathbf{k}_{\mathrm{Tx}}(t)}\, \mathrm{d}t \qquad (2.6)$$

where M_0 is the equilibrium magnetization. For spatial encoding of the magnetization field, the waveform is weighted with a phase term $\mathbf{r}\cdot\mathbf{k}_{\mathrm{Tx}}(t)$. This phase term is controlled by the trajectory $\mathbf{k}_{\mathrm{Tx}}(t)$ in the *transmit k-space* during RF excitation and can be steered by the gradient field $\mathbf{G}(t) = (G_x(t),\, G_y(t),\, G_z(t))$ as:

$$\mathbf{k}_{\mathrm{Tx}}(t) = -\gamma \cdot \int_t^T \mathbf{G}(\tau)\, \mathrm{d}\tau \qquad (2.7)$$

For a typical slice-selective excitation, the transmit k-space trajectory $\mathbf{k}_{\mathrm{Tx}}(t)$ is usually defined by a constant gradient and is applied in conjunction with a sinc-shaped RF pulse (see Fig. 2.1). The corresponding trajectory $\mathbf{k}_{\mathrm{Tx}}(t)$ is a straight line in the transmit k-space traversed at constant velocity, requiring an excitation period in the order of $1\,\mathrm{ms}$ for typical gradient and RF power limits.

The resulting MR signal $s(t)$ induced in the RF receive coils is then the volume integral of the transverse magnetization. When using a quadrature receive (Rx) coil, the signal is observed as rotating in the opposite direction and the B_1^- component acts as the receive sensitivity:

$$s(t) = \text{const} \cdot \int_V M_{xy}(\mathbf{r}) \cdot B_1^-(\mathbf{r}) \cdot e^{i\mathbf{r} \cdot \mathbf{k}_{\text{Rx}}(t)} \, dV \qquad (2.8)$$

During reception, another gradient field $\mathbf{G}(t)$ is applied to introduce the phase term $\mathbf{r} \cdot \mathbf{k}_{\text{Rx}}(t)$, where $\mathbf{k}_{\text{Rx}}(t)$ is the trajectory in the *sampling k-space* [10]:

$$\mathbf{k}_{\text{Rx}}(t) = -\gamma \cdot \int_0^t \mathbf{G}(\tau) \, d\tau \qquad (2.9)$$

Repeating this sequence with different profiles in the receive k-space allows spatial encoding of the magnetization $M_{xy}(\mathbf{r})$ within the scan volume V for reconstruction of an MR image.

2.1.2 Parallel RF transmission

Non-uniformity at high-field MRI

At low field strengths, the RF wavelength is large compared to the dimensions of the human body and the RF transmit field B_1^+ can be considered as spatially uniform. However, already at 1.5 T minor penetration effects due to the reduced wavelength can be observed [1]. At 3 T, the RF wavelength in muscle tissue amounts $\lambda \approx 25$ cm, which is about the diameter of the human body. In this regime, the RF waves entering the human body from different directions experience different phase delays. As a consequence, the resulting transmit field B_1^+ exhibits constructive or destructive interference, depending on the location in the body. This non-uniform B_1^+ field leads to spatially varying magnetization (according to Eq. 2.6), which is a serious problem in diagnostic imaging in high-field MRI, as image contrast may exhibit severe spatial variation. Similarly, the non-uniform receive field B_1^- leads to spatially varying signal intensity, according to Eq. 2.8. These effects might hide pathologies of a patient and can hence mitigate the diagnostic value of high-field MRI.

A linear system perspective

The use of multiple independent Tx coils has been proposed to improve the spatial homogeneity of the RF field [1, 2]. Due to the linearity of Maxwell's equations (see Appendix A.1), the resulting transmit field

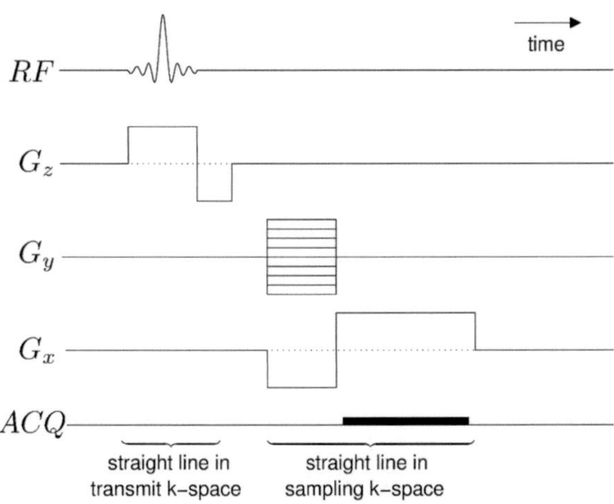

Fig. 2.1: Sequence diagram of a slice-selective, Cartesian gradient echo MR scan: The spatial encoding of the magnetization during RF excitation is attributed to the transmit k-space. In this simple example, a sinc-shaped RF pulse is applied in combination with a constant gradient G_z for slice selection. The slice-select gradient is followed by a negative rephase lobe. Spatial encoding of the magnetization is performed in the so-called sampling k-space, which is defined as a 2D trajectory by the associated gradients G_x and G_y. To allow spatial reconstruction of the full slice, this sequence is performed repeatedly using the same transmit k-space trajectory but different phase-encoding gradients G_y to cover the full sampling k-space.

$B_1^+(\mathbf{r}, t)$ in time t is the linear superposition of the magnetic transmit field sensitivities $\mathbf{S}_{B1}(\mathbf{r})$ of the individual transmit channels:

$$B_1^+(\mathbf{r}, t) = \mathbf{S}_{B1}(\mathbf{r}) \cdot \mathbf{w}(t) = \begin{pmatrix} S_{B1,1}(\mathbf{r}) & \cdots & S_{B1,N}(\mathbf{r}) \end{pmatrix} \cdot \begin{pmatrix} w_1(t) \\ \vdots \\ w_N(t) \end{pmatrix}$$

$$(2.10)$$

where $\mathbf{w}(t)$ denotes a dimensionless vector containing the RF waveforms of each channel.* N is the number of transmit channels. The mag-

---

*Note that B_1 and E are in phasor notation and that the time t refers to the RF waveform which changes much slower than the Larmor rate.

netic transmit field sensitivities $\mathbf{S}_{B1}(\mathbf{r})$ depend on the coil design and the dielectric load (i.e. the patient body shape and position) and can be measured using B_1 mapping techniques (see e.g. [11] for an overview). Due to linearity, the concomitant electric field $\mathbf{E}(\mathbf{r},t)$ can likewise be calculated by a linear superposition of the associated electric field sensitivities $\mathbf{S}_E(\mathbf{r})$:

$$\mathbf{E}(\mathbf{r},t) = \mathbf{S}_E(\mathbf{r}) \cdot \mathbf{w}(t) = \begin{pmatrix} S_{Ex,1}(\mathbf{r}) & \cdots & S_{Ex,N}(\mathbf{r}) \\ S_{Ey,1}(\mathbf{r}) & \cdots & S_{Ey,N}(\mathbf{r}) \\ S_{Ez,1}(\mathbf{r}) & \cdots & S_{Ez,N}(\mathbf{r}) \end{pmatrix} \cdot \begin{pmatrix} w_1(t) \\ \vdots \\ w_N(t) \end{pmatrix}$$

$$(2.11)$$

Unfortunately, the electric field sensitivities inside the human body are not directly accessible by measurement and have to be estimated from numerical simulations, as discussed later in this work.

RF shimming

In RF shimming, the same RF waveform is applied to all Tx channels such that the waveforms $\mathbf{w}(t)$ can be separated into the scalar waveform $a(t)$ and the so-called channel weights $\mathbf{w} = (w_1, \ldots, w_N)^T$:

$$\mathbf{w}(t) = \mathbf{w} \cdot a(t) \tag{2.12}$$

Optimizing the channel weights $\mathbf{w}$ allows to improve the effective transmit field $B_1^+(\mathbf{r})$ for uniform image contrast and intensity (see Fig. 2.2). The term *RF shimming* has been coined for this approach in analogy to *main field shimming*, where small iron pieces (shims) are placed at different locations in the magnet to improve the B_0 homogeneity [12]. Using the same waveform $a(t)$ on all channels has the advantage that RF shimming can be directly applied to any conventional MRI sequence.

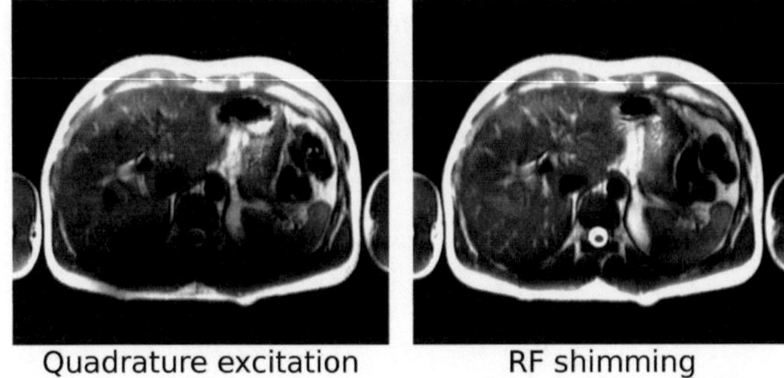

Fig. 2.2: RF shimming in T2-weighted TSE imaging: At conventional quadrature excitation, dark shading can be observed at the posterior side of the body. By RF shimming, using an 8-channel body coil, the wave propagation effects can be compensated. The overall appearance of the image appears more uniform.

Transmit SENSE

The transmit k-space concept offers a much more powerful tool, enabling the excitation of arbitrary spatial magnetization patterns [9]. This can be applied for volume selective excitation and outer volume suppression [13] or for localized motion detection by navigators [14]. This framework, however, requires full coverage of the transmit k-space leading to excitation periods on the order of 10 to 30 ms, hampering practical applications.

Instead of keeping the channel weights fixed during the RF pulse, different RF waveforms $\mathbf{w}(t) = (w_1(t), \ldots, w_N(t))^T$ are applied to each individual RF channel. This approach yields additional freedom to manipulate the transverse magnetization and Eq. 2.6 can be expanded to:

$$M_{xy}(\mathbf{r}) = i\gamma M_0 \mathbf{S}_{B1}(\mathbf{r}) \cdot \int_0^T \mathbf{w}(t) \cdot e^{i\mathbf{r}\mathbf{k}_{\mathrm{Tx}}(t)} \, \mathrm{d}t \qquad (2.13)$$

This freedom can be applied for spatially selective excitation with undersampling in the transmit k-space, accelerating the required pulse to a few milliseconds [3–5]. This concept is referred to as *Transmit SENSE*, in analogy to *sensitivity encoding* (SENSE) [15] for parallel reception which allows undersampling in the sampling k-space for reduced acquisition

times. One application of such accelerated spatially selective excitation
is the acquisition of FFE *zoom images* as shown in Fig. 2.3.

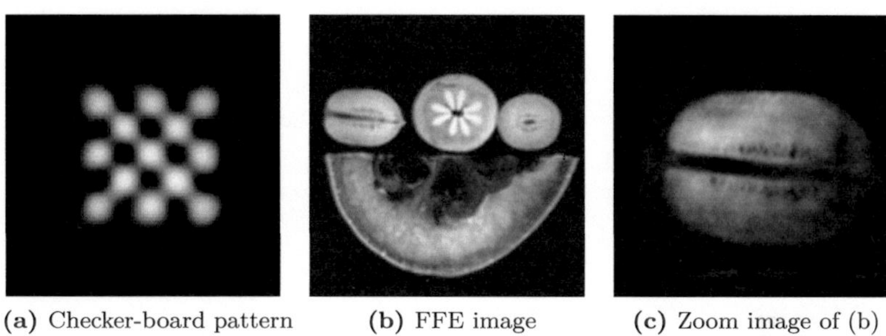

(a) Checker-board pattern   (b) FFE image   (c) Zoom image of (b)

Fig. 2.3: Spatially selective excitation using Transmit SENSE: a) 2D selective excitation of a checkerboard pattern in a homogeneous phantom.
b) Conventional FFE image of some fruits. c) Spatially selective excitation of the kiwi fruit on the upper left. In *zoom imaging*, this is exploited
to accelerate the acquisition by reducing the field-of-view without aliasing
artifacts. *Courtesy of Ingmar Gräßlin.*

2.2 Effects of SAR and temperature in MRI

2.2.1 SAR definition

According to the laws of electrodynamics, the B_1 field used to manipulate
the magnetization of the spin system induces a concomitant E field, and
hence eddy currents, in the body tissues. Due to resistive losses, most
of the RF power transmitted during an MR examination is dissipated
as heat in the patient body [16]. This RF induced heating may lead to
tissue damage (and even burns) as well as to physiologic effects, including
heat sensation and cardiovascular stress. Hence, the applied RF power
needs to be limited to ensure patient safety.

Temperature rise within the human body depends on various factors,
including heat diffusion and bioresponses, and is very difficult to predict.
The local *specific absorption rate* (SAR) [17] is used as a substitute figure
and is defined as the dissipated RF power P_{diss} per mass m of body tissue
in a local volume V (containing e.g. 10 g of tissue):

$$\mathrm{SAR}(\mathbf{r}) = \frac{1}{V} \int_V \frac{P_{diss}}{m} \, \mathrm{d}V = \frac{1}{V} \int_V \frac{\sigma(\mathbf{r})}{2\rho(\mathbf{r})} |\mathbf{E}(\mathbf{r})|^2 \, \mathrm{d}V \qquad (2.14)$$

Here, σ and ρ denote electrical conductivity and mass density of the tissue, respectively, and $|\mathbf{E}(\mathbf{r})|$ is the amplitude of the electric field vector at a location $\mathbf{r}$ in the body.* In addition to the local SAR, the body-averaged SAR is defined as:

$$\mathrm{SAR}_{\mathrm{body}} = \frac{1}{V_{\mathrm{body}}} \int_{\mathrm{body}} \mathrm{SAR}(\mathbf{r}) \, \mathrm{d}V \qquad (2.15)$$

Similarly, the head-averaged SAR and a partial-body SAR can be defined as averages over selected body regions. At thermal equilibrium, the temperature rise is linearly related to the dissipated RF power via the specific heat capacity c_H:

$$\mathrm{SAR} = c_H \cdot \frac{\mathrm{d}T}{\mathrm{d}t} \qquad (2.16)$$

2.2.2 General considerations

Contributing E field components

According to Maxwell's equations (cf. Appendix A.1), an electric field E can arise from two independent sources:

Conservative E fields arise from accumulation of free electrical charges ρ_{el}, located at boundaries with different permittivity, according to Gauss's law. The associated electric field lines enter and exit the body at different locations. In a resonant RF coil structure, the capacitive E field occurs highly localized at the capacitors. This E field is usually minimized in the RF coil design process by using multiple capacitors and distributing them evenly along the coil.

Magnetically induced E fields with closed loop field lines (curls) inside the patient body are caused by the time-varying magnetic field B_1 according to Faraday's law. The corresponding induced

---

*Alternatively, Eq. 2.14 is sometimes written without the factor of 2 in the denominator, referring to the root-mean-square value of the electric field.

currents are referred to as RF eddy currents, flowing inside the body along the E field loops. The magnetically induced E field rises linearly with frequency, and is generally expected to be dominant in human MRI while the conservative E field is often negligible [6, 16, 18].

In close proximity to the capacitors (a few centimeters), the conservative E field needs however to be considered for safety assessment. For parallel transmit MRI coils, each Tx element can be treated separately, if the distance to neighboring Tx elements is large enough (w.r.t. the conservative E field component) and if mutual coupling between the coils is prevented. This implies, that the conventional techniques for design and safety assessment (e.g. by simulations and local temperature measurements) are sufficient to evaluate the safety of the conservative E field.

By contrast, the magnetically induced E field occurs in all body parts exposed to the RF transmit field. In parallel transmit MRI, this E field is equally constituted by all Tx elements. The resulting E field and hence the SAR depends on the transmit phases and amplitudes of all channels. This represents a novel situation compared to conventional MRI systems and is hence the major cause of safety concerns. Consequently, the magnetically induced E field is in the focus of investigation in this work.

Frequency-dependency of the SAR

The SAR increases significantly with the frequency of the applied RF field [7, 18, 19]. This increase can be attributed to the following effects:

Farady's law: The electric field E along a conduction loop (e.g. of conductive tissue in the human body) increases linearly with the time derivative $i\omega B_1$ of the magnetic field through that loop, according to Farady's law. For a given RF field, the SAR hence increases quadratically with the frequency ω, as shown analytically in [7] for a spherical phantom.

Dielectric dispersion: The dielectric tissues properties vary with frequency, due to interaction of the RF field with the cellular and molecular constituents of tissue. This effect is referred to as *dielectric dispersion* [20, 21]. In the RF frequency range, neither the polarization effects at cellular walls (β-dispersion at 10–100 kHz) nor the polarization of water molecules (γ-dispersion in the GHz-region) are very pronounced. Still, conductivity values of human tissues generally increase with frequency in the RF range. For example, the conductivity σ of liver tissue rises by approximately 15% from 64 MHz to 128 MHz. According to Eq. 2.14, this increase also applies to the SAR.

RF field non-uniformity: In high-field MRI (≥ 3 T), the magnetic field B_1 becomes spatially non-uniform in the human body due to interference effects. The induced E field follows from the spatial derivative, more precisely the curl, of the B_1 field according to Ampere's law. A spatially more inhomogeneous B_1 field is hence likely to produce an increased E field at some locations [22].

Decreased quadrature efficiency: RF transmit coils for MRI are usually driven at quadrature excitation. At 64 MHz, this results in an almost ideally circularly polarized transmit field B_1^+, whereas the counter-rotating component B_1^- is close to zero. The associated power demand is one-half of that required for a linearly polarized field [8].

At high fields, the RF waves entering the human body from different sides possess a spatially varying phase delay and the efficiency of the quadrature drive is reduced. At 128 MHz, the average B_1^- field can amount up to about 40% of the B_1^+ field at quadrature excitation, increasing with patient size [23]. Hence, the efficiency of the quadrature drive is significantly reduced and a higher RF power is needed to achieve a desired B_1^+.

In summary, the SAR increase due to Faraday's law and dielectric dispersion arise from fundamental physics and properties of biological tissue which cannot be controlled by technical means. On the contrary, the uniformity and quadrature efficiency of the RF field are determined not only by the patient, but likewise by the design of the applied RF field. This offers a potential for SAR reduction by suitable RF coil design and spatial control of the RF field by parallel transmission.

2.2.3 SAR standards and regulations

Different international safety standards have been developed to ensure that RF heating in MRI stays within safe limits. The major applicable standard is issued by the International Electrotechnical Commission (IEC) [17], defining limits for the SAR based on proposals made by the International Commission on Non-Ionizing Radiation Protection (ICNIRP) [24] and the IEEE [25]. The IEC limits are now also accepted by the U.S. Food and Drug Administration (FDA). The IEC standard defines three operating modes:

Normal Mode: Mode of operation that causes no physiological stress to patients. This mode is routinely used and also applied for patients with heart diseases or disorders in their thermoregulatory system, pregnant women, and infants. This mode demands a main field strength $B_0 \leq 3\,\mathrm{T}$. The global SAR limits are intended to ensure a body core temperature of 39°C or less.

First Level Controlled Mode: Mode of operation that may cause physiological stress to patients and needs to be controlled by medical supervision. The MR scanner will inform the operating person to ensure this. The required main field strength is $B_0 \leq 4\,\mathrm{T}$. The global SAR limits are intended to ensure a body core temperature of 40°C or less.

Second Level Controlled Mode: Mode of operation that can produce significant risk for patients and requires explicit ethical approval of an institutional ethics committee.

The IEC differentiates between volume RF transmit coils, producing a homogeneous Tx field, and local RF transmit coils, comprising any other RF transmit coils. For volume RF transmit coils, the IEC specifies limits for the global SAR, averaged over the whole body, the head, or the exposed body region. For local RF transmit coils, additional limits are defined for the local SAR, averaged over 10 g of tissue. Parallel transmit coil arrays can function either as volume RF transmit coils or as local RF transmit coils. RF shimming aims to generate a homogeneous Tx field, hence a volume coil might be an appropriate classification. However, the contributions of the individual Tx channels can significantly deviate from each other, depending on the application. It is hence prudent to consider the local SAR in any case.

The SAR limits according to the IEC are summarized in Tab. 2.1. For comparison, the basal metabolic rate is about 1.3 W/kg without exercise and can reach up to 18 W/kg for athletes during intensive exercise [26,

27]. The IEC regulations also require an ambient room temperature of 25°C or less. For higher room temperatures, reduced SAR limits are specified.* SAR values are generally averaged over a six-minute time window. Additionally, the maximum admissible energy deposited during an examination is limited to 14,400 Ws/kg (corresponding to 1 hour at the maximum global SAR in IEC first mode).

Tab. 2.1: SAR limits for the different operating modes according to the IEC standard.

	Global body SAR	Global head SAR	Local head and torso SAR	Local extremity SAR
IEC Normal mode	2 W/kg	3.2 W/kg	10 W/kg	20 W/kg
IEC First level controlled mode	4 W/kg	3.2 W/kg	20 W/kg	40 W/kg
IEC Second level controlled mode	>4 W/kg	>3.2 W/kg	>20 W/kg	>40 W/kg

2.2.4 Global SAR concerns

The global (body averaged) SAR can result in an increase of the patient's body core temperature [28]. A healthy body will counteract heating by thermoregulatory responses, including an increased respiratory rate and an increased heart rate combined with widening of the arterioles to increase blood flow. Another response is whole-body sweating [29], evaporation then transfers thermal energy to the surrounding air.

The global SAR is hence of special concern for patient groups with limited thermoregulatory capabilities. These include neonates and elderly patients, sedated and anesthetized patients, pregnant women and patients with diabetes [30]. It is hence at the discretion of the clinicians whether scanning in normal mode or first level controlled mode is advisable.

---

*Until March 2010, when the 3rd edition of the IEC standard was released, the room temperature limit was at 24°C and additional restrictions on room humidity were applicable.

Technologically, the global SAR can be measured relatively directly. This can be achieved either by measurement of the temperature rise in a thermally isolated phantom (calorimetric method) or by monitoring of the forward and reflected RF power (pulse energy method). Both methods are described in a standard of the National Electric Manufacturers Association (NEMA) [31]. Recently, the pulse energy method has been extended for multiple feeding ports to allow measurement of the global SAR for parallel transmit MRI [32].

2.2.5 Local SAR concerns

Localized heating caused by local SAR hotspots is a further point of concern in RF transmission during MRI. Human skin begins to redden after about 2 minutes at 45°C when proteins start to denature, whereas full thickness burns appear after about 100 minutes [26]. Furthermore, medical implants and interventional medical devices may focus the E fields and significantly increase local temperature [33]. Safety assessment of such devices is an extensive field of research [16], but beyond the scope of this work.

Local SAR hotspots can arise from the conservative E field in proximity to the RF coils as well as from the magnetically induced E field at narrow points of the eddy current pathways inside the human body. Such narrow points can occur either inside the patient body [18], depending on the local tissue composition, or from closed loops formed by the body positioning, e.g. by hands touching the body or touching thighs [26]. To prevent such loops from body positioning, it is good clinical practice to avoid contact of the extremities by inserting non-conducting pads.

Thanks to such preventive measures, the number of reported incidents related to RF transmission is relatively small. Reported incidents are mostly associated with a high SAR scan and often involve skin-to-skin contact or body contact to the RF coil or to foreign objects [30,34]. Resulting thermal injuries were primarily located at the arms and the legs, but also occurred at the torso and shoulders [34].

Unlike the global SAR, the local SAR is not directly accessible by measurement. Only recently, Katscher *et al.* [35] proposed a method to estimate the local SAR from measured B_1^+ maps. Todate, this approach is however limited to subregions (typically 2D planes) of the patient body. In practice, local SAR estimates need to be obtained based on electro-magnetic (EM) field simulations using discretized human body models. Such simulations have demonstrated that the local SAR limits tend to be reached long before the global SAR limits for typical RF body coils [36–39]. The local SAR values in theses studies varied significantly depending on the patient model used and the body position within the MR scanner. Given the limited number of body models available, investigations are required to fully understand these effects and to extend present studies to parallel transmit MRI.

2.2.6 Temperature simulations and measurements

As temperature is the original cause of concern, several researchers have made attempts to predict temperature rise from a simulated SAR distribution [29,40]. Generalizing Eq. 2.16, the thermal energy balance in perfused body tissue can be described by Pennes' bioheat equation:

$$\rho \cdot c_H \cdot \frac{\mathrm{d}T}{\mathrm{d}t} = \nabla(k_H \nabla T) - w \rho_{\mathrm{blood}} c_{\mathrm{blood}} (T - T_{\mathrm{blood}}) + q_m + \rho \cdot \mathrm{SAR} \quad (2.17)$$

The first term on the right-hand side describes the heat diffusion, with k_H being the thermal conductivity. The second term describes the heat transfer due to blood perfusion, where w is the perfusion rate, ρ_{blood} and c_{blood} are the density and specific heat capacity of blood, and T_{blood} is the basal blood temperature. The heat production rate of the metabolism q_m represents the internal heat source of the body. To model RF heating in MRI, the SAR term is added as an external heat source.

Temperature modeling is a challenging task, as the actual temperature rise in the human body also depends on additional physiological parameters. Some authors also included thermoregulatory responses [41] as well as the detailed vascular network [42]. Since these contributions are highly patient-dependent and difficult to model, the use of temperature simulations instead of the SAR would require very conservative assumptions. Due to such uncertainties, this approach is often more restrictive than using the current SAR limits [40] and hence would likely not be accepted as a replacement of the well-established SAR limits. Temperature simulations are hence not pursued in this work.

In vivo temperature measurements via MRI are in principle feasible by various approaches (see [43] for a review). The proton resonance frequency (PRF) shift method is a relatively sensitive and highly linear approach to measure temperature changes. The PRF method is hence the preferred choice for many applications, including high-intensity focused ultrasound (HIFU) applications [44]. Recently, simulations of RF heating in the human forearm could successfully be validated by the PRF method [45]. The *in vivo* sensitivity of the PRF method can range down to 0.5 K, if sufficient imaging time is spent. Imaging at such a high temperature resolution however conflicts with the clinical protocols. Thus, MR temperature measurements can at present not be considered as suitable for monitoring of RF heating during MR imaging.

2.3 SAR calculation

2.3.1 SAR calculation for conventional MR sequences

Based on EM simulations using a human body model, the electric fields $E(\mathbf{r})$ can be predicted. Eq. 2.14 describes the SAR for a single point in time. To obtain an expression for the average SAR of an MR scan, a "duty cycle" factor a_{rms} is introduced and defined as the root mean square value of the RF waveform $a(t)$, averaged over the entire scan (i.e. including all information about repetition times):

$$a_{\mathrm{rms}} = \sqrt{\frac{1}{T} \int_{\mathrm{scan}} |a(t)|^2 \, \mathrm{d}t} \tag{2.18}$$

The local SAR can then be calculated as:

$$\mathrm{SAR}(\mathbf{r}) = \underbrace{\frac{1}{V} \int_V \frac{\sigma(\mathbf{r})}{2\rho(\mathbf{r})} |\mathbf{E}(\mathbf{r})|^2 \, \mathrm{d}V}_{q(\mathbf{r})} \cdot a_{\mathrm{rms}}^2 \tag{2.19}$$

The scalar constant $q(\mathbf{r})$ denotes the normalized SAR value at each location in the simulated body model. For calculation of the body- and head-averaged SAR, two similar constants q_{body} and q_{head} can be obtained by averaging over the entire body model or the head, respectively (cf. Eq. 2.15).

For single-channel RF transmission, the SAR hotspot locations do not change for given models of the RF coil and the patient body. Hence, only the maximum values of $q(\mathbf{r})$ at the hotspots in the head, torso, and extremities are needed to calculate the maximum local SAR in the patient. For each transmit coil (e.g. the Tx body coil and different Tx head coils), the appropriate q values are stored at the MR scanner to predict the SAR of an MR sequence.

Such numerical calculations of local SAR values are routinely performed for safety assessment of single-channel RF transmit coils. Usually, several simulations are performed, using different body models in a complete range of body positions within the Tx coil [6].

2.3.2   SAR calculation for RF shimming

When using multiple Tx channels, the global and the local SAR can change significantly, depending on the wave interference of the RF fields transmitted by all channels. The SAR prediction concept has hence to be extended to account for this novel situation. First approaches aimed to estimate the worst-case SAR that can occur for any combination of channel weights [46,47]. However, such approaches significantly overestimate the actual SAR and restrict the allowed RF duty-cycle significantly. This would prohibit even the use of sequences that are allowed with conventional quadrature excitation [48,49] and is hence not practicable. Instead, the actual channel weights $\mathbf{w}$ have to be considered in the SAR prediction.

The so-called "Q-matrix formalism" aims for efficient SAR prediction in human body models. The approach was originally introduced for hyperthermia treatment [50], and allows efficient real-time SAR calculations for parallel transmit MRI [4,48].

Combining Eqs. 2.11 and 2.19 allows to calculate the local SAR for multiple Tx channels as:

$$
\mathrm{SAR}(\mathbf{r}) = \mathbf{w}^H \cdot \underbrace{\frac{1}{V} \int_V \frac{\sigma(\mathbf{r})}{2\rho(\mathbf{r})} \mathbf{S}_E^H(\mathbf{r}) \cdot \mathbf{S}_E(\mathbf{r}) \, \mathrm{d}V}_{\mathbf{Q}(\mathbf{r})} \cdot \mathbf{w} \cdot a_{\mathrm{rms}}^2 \quad (2.20)
$$

$$
= \mathbf{w}^H \cdot \mathbf{Q}(\mathbf{r}) \cdot \mathbf{w} \cdot a_{\mathrm{rms}}^2 \quad (2.21)
$$

where the superscript ^{H} denotes the conjugate transpose. In this expression, the integral term is independent of the actual channel weights $\mathbf{w}$ and can be summarized as an $N \times N$ matrix $\mathbf{Q}(\mathbf{r})$ which is Hermitian and positive definite. The term "Q-matrix" was chosen as the matrix represents a quadratic form of the electric field sensitivities. The advantage of this form is that the local Q-matrices for each mesh cell can be pre-calculated from the simulation data, independent of the applied channel weights. Similarly, two additional Q-matrices can be calculated for the global SAR estimation by averaging over the whole body or the whole head.

Furthermore, it is useful to define the root-mean-square value of the effective B_1 field as a reference measure:

$$B_{1,\mathrm{rms}} = < \mathbf{S}_{B1}(\mathbf{r}) \cdot \mathbf{w} > \cdot a_{\mathrm{rms}} \tag{2.22}$$

where the angular brackets denote the spatial average over the field magnitude in a region of interest.

2.3.3 SAR calculation for Transmit SENSE

In the general framework of Transmit SENSE, different RF pulses are applied to each Tx channel. In this case, the channel weights $\mathbf{w}$ in Eq. 2.20 are replaced by the RF waveforms $\mathbf{w}(t) = (w_1(t), \ldots, w_N(t))^T$. The expression for the SAR then becomes:

$$\mathrm{SAR}(\mathbf{r}) \quad = \quad \frac{1}{T_{\mathrm{scan}}} \int_{\mathrm{scan}} \mathbf{w}^H(t) \cdot \mathbf{Q}(\mathbf{r}) \cdot \mathbf{w}(t) \, \mathrm{d}t \tag{2.23}$$

Note, that the "duty cycle" factor a_{rms} is no longer needed, as the time signal is now represented by $\mathbf{w}(t)$. In this equation, a matrix operation has to be performed for every point in time, which is computationally not efficient. Instead, a separation of the temporal and spatial components is proposed here. Starting with a discretized form of Eq. 2.23 into N_T sample points, the expression can be reformulated as:

$$\mathrm{SAR}(\mathbf{r}) \quad = \quad \frac{1}{N_T} \sum_{k=1}^{N_T} \mathbf{w}^H(t_k) \cdot \mathbf{Q}(\mathbf{r}) \cdot \mathbf{w}(t_k) \tag{2.24}$$

$$= \quad \sum_{m=1}^{N} \sum_{n=1}^{N} Q_{m,n}(\mathbf{r}) \cdot \underbrace{\frac{1}{N_T} \sum_{k=1}^{N_T} w_m^*(t_k) \cdot w_n(t_k)}_{} \tag{2.25}$$

$$= \quad \sum_{m=1}^{N} \sum_{n=1}^{N} Q_{m,n}(\mathbf{r}) \cdot \quad \mathrm{E}\big[w_m^*(t_k) \cdot w_n(t_k)\big] \tag{2.26}$$

The operator $\mathrm{E}[.]$ denotes the expected value (mean) and is applied to calculate the average cross-power of every two waveforms $w_m(t)$ and

$w_n(t)$. This expression is independent of the spatial location and can be pre-calculated. The SAR can then be efficiently calculated for each location $\mathbf{r}$ by a weighted double-sum over all elements $Q_{m,n}$ of the matrix $\mathbf{Q}$.

2.3.4 Q-matrix model compression

The Q-matrix formalism yields one matrix $\mathbf{Q}(\mathbf{r}_i)$ for each mesh cell i of the simulated body model (i.e. approx. 400.000 matrices per model). Though it is computationally feasible to calculate the maximum local SAR in real-time [48], it is desirable to reduce this huge amount of cells to a smaller number of so-called observation points to reduce memory requirements and increase computational efficiency.

Such an observation point j should represent an upper bound to the local SAR at another point i, which could then be excluded from the calculation of the maximum local SAR:

$$\mathrm{SAR}(\mathbf{r}_i) \;\leq\; \mathrm{SAR}(\mathbf{r}_j) \qquad \forall \mathbf{w} \tag{2.27}$$

$$\Leftrightarrow\; \mathbf{w}^H \mathbf{Q}(\mathbf{r}_i)\mathbf{w} \;\leq\; \mathbf{w}^H \mathbf{Q}(\mathbf{r}_j)\mathbf{w} \;\; \forall \mathbf{w} \tag{2.28}$$

Unfortunately, Eq. 2.27 is not fulfilled for most pairs of mesh cells because $\mathrm{SAR}(\mathbf{r}_j)$ can get down to almost zero for some channel weights (i.e. the smallest eigenvalue of $\mathbf{Q}(\mathbf{r}_j)$ is close to zero). Gebhardt *et al.* [51,52] introduced an additive SAR overestimation term ε to address this problem. The augmented problem is given by:

$$\mathbf{w}^H \mathbf{Q}(\mathbf{r}_i)\mathbf{w} \;\leq\; \mathbf{w}^H \left(\mathbf{Q}(\mathbf{r}_j) + \varepsilon \mathbf{I}\right)\mathbf{w} \qquad \forall \mathbf{w} \tag{2.29}$$

$$\Leftrightarrow \qquad -\varepsilon \;\leq\; \lambda_{\min}[\mathbf{Q}(\mathbf{r}_j) - \mathbf{Q}(\mathbf{r}_i)] \tag{2.30}$$

where $\mathbf{I}$ denotes the identity matrix. Whether the mesh cell i can now be excluded from the SAR calculation can be decided by comparing ε with the smallest eigenvalue $\lambda_{\min}$ of $\mathbf{Q}(\mathbf{r}_j) - \mathbf{Q}(\mathbf{r}_i)$.

Finally, all remaining matrices $\mathbf{Q}_j$ are replaced with the augmented matrices $\mathbf{Q}_j + \varepsilon \mathbf{I}$ to ensure that the SAR calculated with the compressed model is an upper bound to the SAR of the uncompressed model.

2.4 Conclusions

RF parallel transmission offers a significant potential to enhance image uniformity and overall RF excitation performance. Furthermore, an appropriate control of the Tx coil array allows to counteract the SAR increase in high-field MRI. Whereas the magnetic transmit field can be determined by measurement, the concomitant electric fields are more difficult to predict. The global SAR is accessible via power or temperature measurements, however the local SAR cannot be measured in clinical practice and tends to be more limiting with respect to current safety regulations. To date, EM simulations represent the only feasible solution for local SAR prediction.

Numerical recipes for efficient SAR calculations, based on the electric fields obtained from numerical simulations, are available and have been summarized in this chapter. It is, however, challenging to realize these ideas by suitable modeling of parallel transmit RF coils and, more importantly, of different volunteers for a valid SAR prediction. These issues are addressed in the following chapters of this work.

Chapter 3

Modeling of the Multix Body Coil

For numerical SAR simulations, adequate models of the RF coil as well as of the patient body are required. This chapter discusses the generation of such a model for the eight-channel Multi-Transmit Body Coil (MBC) [53]. The generation of suitable human body models will be covered later in Chapter 4. The MBC is fully integrated in a 3 T MR scanner (Achieva, Philips Medical Systems, Best, The Netherlands) which is extended to eight Tx channels such that each Tx element can be controlled independently [54]. The system is located at Philips Research Hamburg and was used for all experiments in this work.

Over the past decades, a broad variety of numerical methods has been developed for electro-magnetic (EM) field simulations. These include the Finite Elements Method, Finite Difference Methods, and the Method of Moments [55]. Simulation of an RF field inside the human body requires highly detailed models of the human body and its dielectric tissue properties, resulting in millions of unknowns which is a major restriction for the choice of a numerical method.

The finite-difference time-domain (FDTD) method [56, 57] is based on explicit forward-differences in the time domain. Consequently, memory and time requirements of FDTD are linear in the number of unknown variables. The algorithm is well suited to solve very large numerical systems. By contrast other common numerical methods require a matrix inversion and are hence less efficient when the number of unknowns is very large. Hence, FDTD has gained wide popularity for SAR simulations and is used exclusively in this work. A drawback of FDTD is that all geometric structures are represented on a cubical simulation grid. Especially for thin conductors, this can lead to modelling errors. Hence, care has to be taken when modeling an RF coil as discussed later in this chapter.

3.1 Simulation of the MBC

3.1.1 Description of the MBC hardware

The MBC is composed of eight independent resonator elements which can be used for transmission as well as for reception. As illustrated in Fig. 3.1, these elements are equally spaced circumferentially at 45° inside an RF screen ($\varnothing$675 mm). The elements have different distances to the RF screen, depending on the element location ($d_{1,3,6,8} = 35$ mm, $d_{2,7} = 32$ mm and, $d_{4,5} = 40$ mm). The length of each Tx element is 42 cm and the total width is 10 cm. Each element is designed as a TEM (transversal electro-magnetic) resonator [58] with two parallel strip-lines, as shown in Fig. 3.2a. The strip-lines are connected to the RF screen by the capacitors C_1 and C_2 as indicated in the circuit diagram Fig. 3.2b. In the TEM mode, the currents flow equally through both strip-lines and back through the RF screen ($I_1 = I_2 = I$). Each strip-line is sectioned by four series capacitors C_S which reduces the variation of the current along the element to about 1% [53]. This allows to describe the element with a lumped-component network and also reduces the local conservative E fields at the capacitors. Both strip-lines are connected by two cross-strips, marked L_x, which were introduced to suppress the undesired ring mode ($I_1 = -I_2$) at the RF frequency.

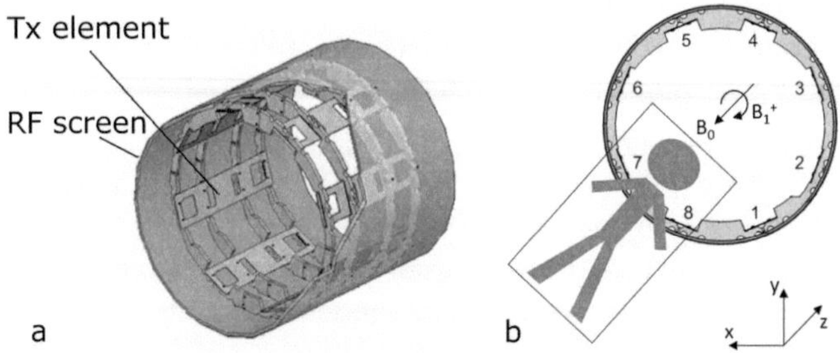

Fig. 3.1: Illustration of the Multix Body Coil (MBC): a) 3D drawing of the MBC with its 8 Tx elements. b) Sketch of the MBC viewed from the side of the patient table: The B_0 field points towards the patient table, which is anti-parallel to the z-axis of the coordinate system used for the FDTD simulations. The patient is shown in a typical head-first scan position.

The capacitors C_M are used for impedance matching to the $50\,\Omega$ of the feeding chain. Furthermore, all neighboring elements are connected by a copper ring and the capacitors C_{D1} and C_{D2} for capacitive decoupling. Each Tx element is equipped with a pick-up coil (PUC) to allow a direct monitoring of the element current. Moreover, the elements can be detuned using PIN diodes to enable the use of other receive and transmit coils.

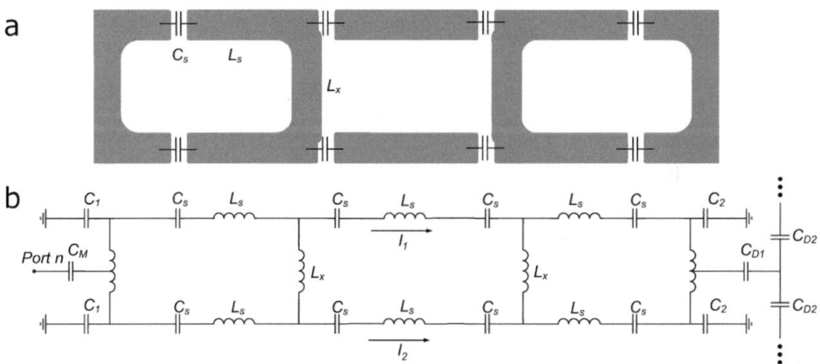

Fig. 3.2: Geometry and circuitry of one Tx element: a) geometric drawing (top view) b) circuit diagram (also showing the tuning capacitors C_1 and C_2, the feeding port and the decoupling network which connects to the neighboring elements). *Courtesy of Peter Vernickel.*

Spatial orientation

To model the RF coil and the patient relative to the main magnet correctly, the spatial orientation needs to be considered (see Fig. 3.1b). In cylindrical Philips MR scanners, the main magnetic field B_0 points out of the bore, towards the patient table. The transmit-active magnetic field component B_1^+ is left-handed with respect to B_0 [6, 7]. On the contrary, B_1^+ is usually defined as rotating mathematically positively (right-handed / counter-clockwise) with respect to the z-axis of the coordinate system [1], as implied in Eq. 2.4. This is also the case for the simulation software used in this work (Remcom XFDTD). This implies that the B_0-axis and the z-axis need to be oriented in an anti-parallel way as shown in Fig. 3.1b. For a typical head-first orientation, the patient's head hence points in positive z-direction.

To generate a circularly polarized transmit field B_1^+, the Tx elements are driven with equal amplitudes and a phase advance of $\phi = n \cdot 45°$, increasing with the element number n. This drive mode is referred to as "quadrature excitation" and mimics the behavior of a birdcage resonator.

3.1.2 Generation of the FDTD model

FDTD geometry model

A simulation model of the MBC was created using the commercial simulation software XFDTD (Remcom Inc., PA, USA). All simulations were performed on a 64-bit desktop PC equipped with two GPU cards (NVIDIA FX5600, each with 1.5 GB memory size and 128 processor cores, memory bandwidth 76.8 GB/sec).

The geometrical structure of the final FDTD model is shown in Fig. 3.3, illustrating the discretization which is discussed further below. The detuning circuitry and the matching network were not included in the simulation. The pick-up-coils were also not included in the FDTD model, instead the element currents are monitored by "point sensors" located at the lumped capacitors for normalization purposes.

Resonance tuning

The capacitors were modeled as lumped elements and each Tx element was tuned to the desired resonance frequency. This tuning can be achieved iteratively by exciting the resonator with a broadband signal (e.g. a Gaussian pulse) and iteratively varying the capacitor values until the desired resonance frequency is reached [59]. A more efficient way is to calculate the loop inductance from a simulation with initial capacitance values and then to calculate the tuned capacity values [60]. Alternatively, current sources can be used to calculate the loop inductance [61].

The aim is to tune the resonance frequency of the TEM mode to 128 MHz. When a Tx element is modeled in an orientation parallel to the 3D FDTD grid (e.g. between elements 1 and 8), the discretized strip-lines are entirely flat and symmetric to each other, such that the resonance frequencies of both strip lines coincide. This results in a distinct first resonance frequency as plotted in Fig. 3.4 (tuned to 128 MHz). Several higher order modes occur above 300 MHz but these are not of interest here.

When the element is rotated by 22.5° as shown in Fig. 3.3b, the discretization of the strip-lines results in stair-casing. As the exact symmetry of the model is lost, both strip-lines now exhibit slightly different impedances. Instead of a single resonance peak, multiple resonance peaks were observed in proximity to the desired resonance. This results from mutual coupling between the two strip-lines. The exact frequencies of these resonance peaks were found to depend on the actual discretization of the strip-lines, i.e. the number and width of the staircases. This effect obviated a steady state solution of the initial simulations.

To avoid this behavior, the discretization of the strip-lines of the rotated Tx elements were manually re-adjusted to achieve the same staircase-pattern for both strips. After this refinement, the resonance modes (see Fig. 3.4) were separated by at least 7 MHz. An energy transfer between the two strip-lines was then no longer observed and the simulations converged to a steady-state. Comparison of the currents in all capacitors showed equal phases and amplitudes (not varying by more than 5%), demonstrating that the selected resonance mode is the correct TEM mode.

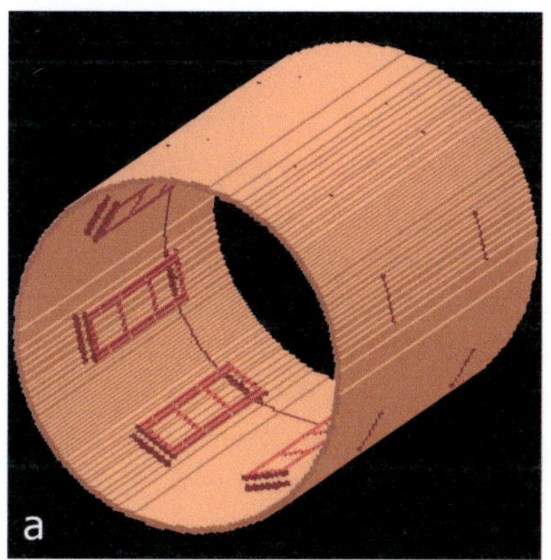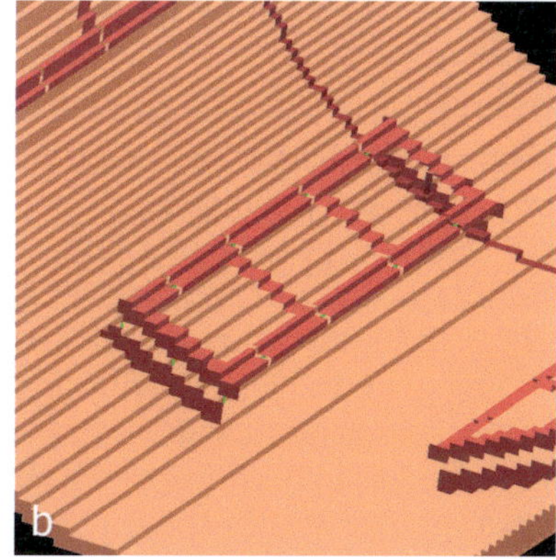

Fig. 3.3: Simulation model of the MBC at 5 mm grid resolution: a) Complete MBC, b) Detailed view of one TEM element, each strip-line is equally discretized by two stairs.

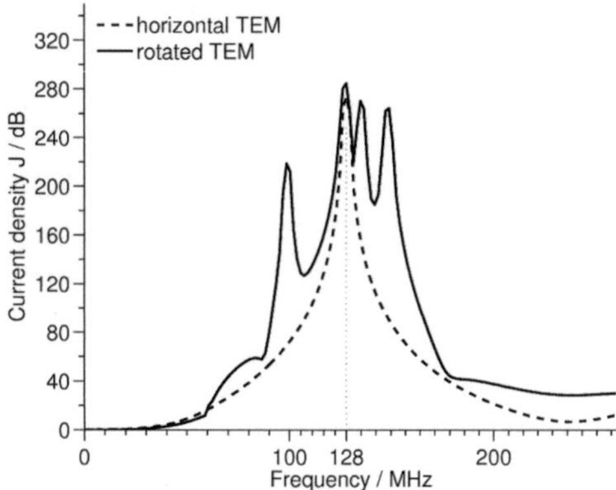

Fig. 3.4: Frequency responses of a Tx element: When placed in parallel to the FDTD grid (horizontal TEM), a distinct first resonance mode occurs which was tuned to 128 MHz. After rotating the Tx element by 22.5°, multiple resonance modes were observed. After ensuring that both strip-lines are equally discretized, these modes were separated by at least 7 MHz. The resonance mode with the highest amplitude could be verified to be the desired TEM mode and was also tuned to 128 MHz.

Distance to the RF screen

Another discretization issue is that the different distances of the Tx elements to the RF screen cannot be modelled realistically using a grid size of several millimeters. To investigate the effect of the distance to the RF screen, a Tx element was simulated at distances of 35 mm and 40 mm, using a 5 mm grid spacing. The simulation results were then scaled to $B_1^+ = 1\,\mu\text{T}$ in the isocenter of the MBC. After this normalization, the E field and B_1 field distribution showed to be practically independent of the distance to the RF screen. Quantitatively, the difference was less than 1% at any location of the patient space. Significant deviations were however observed in the gap between the element and the RF screen.

In summary, this indicates that both representations of the Tx elements are equally acceptable. However, a normalization is required to obtain meaningful results. Such a normalization will be introduced in the next paragraph.

3.1.3   Calibration of the simulation and the hardware

A correspondence between the simulation and the physical MR hardware needs to be established by a cross-calibration. A broad variety of Tx arrays for parallel transmission exists, varying in load-sensitivity, strength of inter-element coupling, and presence of pick-up coils (PUCs). Depending on these parameters, a suitable calibration needs to be designed.

Normalization

Power-based calibration: The commonly used approach to normalization is to measure the RF forward power at the physical coil and to scale the simulated forward power to that value [46]. In many cases, it is however difficult to model the relation between the forward power and the resulting RF field correctly, as the relation depends on multiple patient-dependent factors (i.e. power reflection, coil resonance quality, and inter-element coupling). One approach to model these effects comprehensively is to combine the EM simulation with a port-based post-processing [62]. In this approach, the S-parameter matrix of the loaded coil array is measured and the pre-simulated EM fields are superpositioned such that the same S-parameter matrix is achieved in the simulation.

Element-current-based calibration: If the RF coil is equipped with pick-up coils (PUCs), calibration can alternatively be performed based on measurement of the amplitudes and phases of the currents in the Tx elements. This is motivated by the fact that the RF field generated by one Tx element is proportional to the element current. This represents a more direct access than a power-based calibration, avoiding the load-dependent uncertainties and not requiring a phase-sensitive power measurement in the MR scanner.

Coupling

The magnetic field generated by each Tx element induces opposing currents in the other elements, according to Lenz's law, which is referred to as "coupling" and needs to be considered in the calibration. The element currents can be measured, if pick-up coils are located at each Tx element.

Given the vector of voltages $\mathbf{V}_{\mathrm{DA}}$ at the digital-to-analog converter, the vector of PUC signals $\mathbf{V}_{\mathrm{PUC}}$ can be expressed by the linear equation:

$$\mathbf{V}_{\mathrm{PUC}} = \mathbf{A}_{\mathrm{sys}} \cdot \mathbf{V}_{\mathrm{DA}} \tag{3.1}$$

where $\mathbf{A}_{\mathrm{sys}}$ is referred to as the "system matrix", describing the resonance quality and mutual inductive coupling between the elements [53]. Complete modeling of this relation is again difficult, as $\mathbf{A}_{\mathrm{sys}}$ depends not only on the static RF chain (cable losses, RF amplifier calibration, coil geometry and circuitry) but also on variable, patient-dependent factors (power reflection, coil resonance quality, and inter-element coupling).

Instead, ideal decoupling can easily be enforced in simulations by simulating each element separately while the remaining elements are detuned [63]. The simulation is then run for each Tx element, with the element current I_i in the i-th element being non-zero in the i-th simulation, while all remaining element currents are forced to zero. There are two approaches for linking these ideally decoupled simulations to reality:

Active decoupling: The system is controlled to match with the decoupled simulations. To achieve this, the system matrix $\mathbf{A}_{\mathrm{sys}}$ is determined first, using the pick-up coils to measure the element currents. Then, the inverse system matrix $\mathbf{A}_{\mathrm{sys}}^{-1}$ is calculated and applied to the Tx signals directly before DA conversion to compensate the coupling between the physical Tx elements [64, 65]. This facilitates a relatively direct control of the individual element currents.

Linear superposition of the simulation results: The RF field sensitivities of the coupled system are predicted using the measured system matrix ($\mathbf{S}_{\mathrm{coupled}} = \mathbf{A}_{\mathrm{sys}} \cdot \mathbf{S}_{\mathrm{uncoupled}}$). This approach allows to simulate the coupled system without need to modify the Tx signals.

Implementation

In this work, an element-current-based calibration with active decoupling was used. The simulation results were normalized to an element current that produces $1\mu\mathrm{T}$ in the center of the empty coil simulation. At the physical MR scanner, the pick-up coils of the MBC are designed as non-resonant figure-of-eight coils and hence relatively insensitive to loading and to B_1 fields from neighboring Tx elements. The pick-up coil signals were calibrated to the signal which corresponds to $1\,\mu\mathrm{T}$, as measured with a small oil phantom in the isocenter, using active decoupling.

3.1.4 Simulation results for the empty MBC

For verification, the empty MBC was simulated at quadrature excitation. The resulting magnetic and electric fields are shown in Fig. 3.5. The transmit active field component B_1^+ is highly uniform throughout the scan volume. The reversely polarized component B_1^- is similar to B_1^+ close to the coil elements, but quickly decays to zero towards the patient space. This means that the quadrature efficiency of the simulated empty MBC is almost 100%. Ideally, only transversal magnetic fields B_1^+ and B_1^- are expected, but an axial component B_{1z} is also observed due to the finite length of the coil elements. In the xy-isocenter plane and along the z-axis, the axial field component B_{1z} is close to zero due to symmetry, i.e. magnetic field components generated by opposite structures cancel out.

As for the E field, the E_z component is generally dominant, whereas E_x and E_y are comparably small. There are two reasons for this: First, the conservative E field component occurs mainly in the direction of the capacitors C_S, i.e. in z-direction. Second, the magnetically induced E field component mainly results from B_1^+ and points in z-direction (resulting from the curl operator with $\mathrm{d}B_1^+/\mathrm{d}z \approx 0$ and $B_1^-, B_z \approx 0$). Towards the ends of the bore, transverse E field components are also observed, as B_1^+ decreases ($|\mathrm{d}B_1^+/\mathrm{d}z| \gg 0$). On the contrary, the transverse E field is zero in the coil's center, where the magnetic field is spatially constant.

3.2 Validation for the empty MBC

3.2.1 Experimental validation of the B_1 fields

B_1 mapping protocol

For experimental validation, transverse B_1 maps were acquired for each Tx channel, using a cylindrical oil phantom ($\varnothing 40\,\mathrm{cm}$, thickness $10\,\mathrm{cm}$). The dielectric properties are comparable to vacuum ($\varepsilon_r \approx 3\ldots 4$, $\sigma \approx 0\,\mathrm{S/m}$). Hence, the measured B_1 maps can be validated against B_1

fields obtained from a simulation of the empty MBC. Measurements were carried out with and without active decoupling.

The *actual flip-angle imaging* (AFI) technique is an approach to B_1 mapping in the steady-state of the spin-system and is hence relatively fast [66,67]. The method consists of two interleaved fast field echo (FFE) acquisitions with different repetition times TR_1 and TR_2. The scan parameters used were: $TR_1 = 20\,\mathrm{ms}$, $TR_2 = 100\,\mathrm{ms}$, and $TE = 1.8\,\mathrm{ms}$. A FOV of $480 \times 480\,\mathrm{mm}^2$ and a transverse resolution of $5 \times 5\,\mathrm{mm}^2$ were used. To increase SNR, an all-but-one excitation scheme was used (i.e. eight measurements were performed, driving seven elements and leaving one out each time) [68, 69]. From this set of linear combinations, the B_1 maps of all single elements were calculated.

Postprocessing and results

The region outside the phantom was masked out in the maps and the simulation and all fields were normalized to their mean (shown in Fig. 3.6). When active decoupling is not applied (Fig. 3.6a), the measured fields were asymmetric and a "transmitter blind-spot" can be observed, as the induced currents produce a constructively interfering B_1^+ field on one side and a destructively interfering B_1^+ field on the other. This is a well-known effect of mutual coupling between coil elements [46]. With active decoupling (Fig. 3.6b), the measured B_1 field is much more symmetric and corresponds with the simulation (Fig. 3.6c). Since the difference is hardly visible by eye, the profiles are shown in Fig. 3.7. The progression of the curves shows good qualitative agreement. For quantitative error assessment, the coefficient of variation (CV – defined as the standard deviation divided by the mean) was calculated for all elements and is listed in Tab. 3.1. The average deviation was 9.75%.

coil 1	coil 2	coil 3	coil 4	coil 5	coil 6	coil 7	coil 8	avg.
11.5%	12.7%	9.9%	10.3%	12.1%	6.8%	6.7%	8.1%	9.75%

Tab. 3.1: Coefficients of variation (CV) of the B_1 maps of all Tx elements, calculated for a transverse slice.

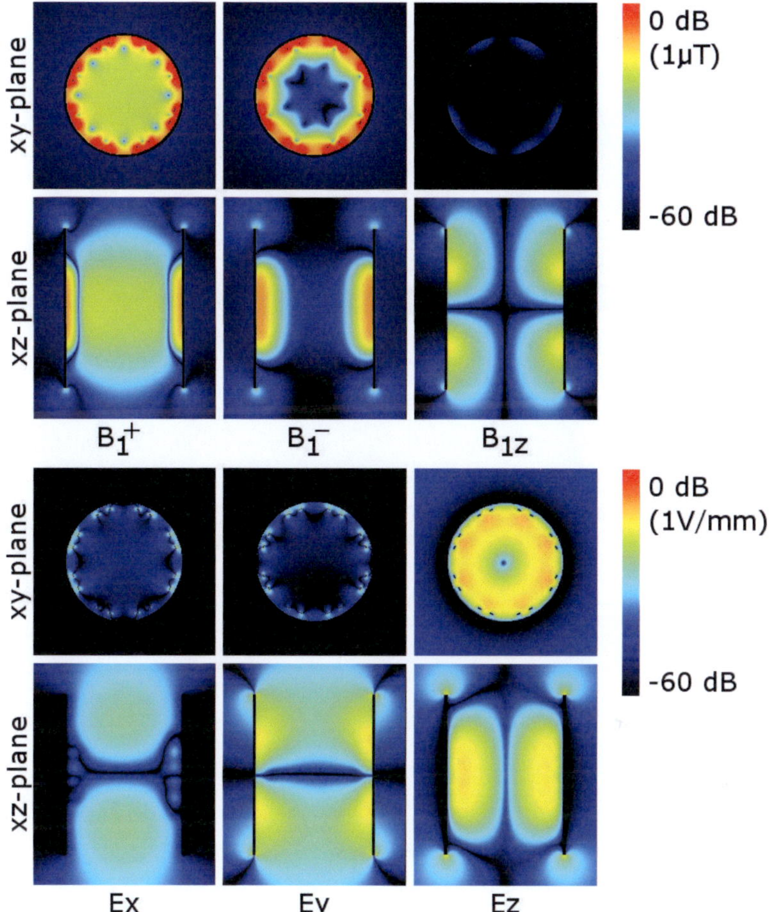

Fig. 3.5: Simulated B_1 and E fields in the empty MBC at quadrature excitation: The transmit-active magnetic field B_1^+ is highly homogeneous, whereas the non-transmit-active component B_1^- quickly tends to zero. Results for the xy-plane are identical to the xz-plane.

The E_z component along the Tx elements is generally dominating. Only towards the ends of the bore where B_1 strongly changes locally, noteable E_x and E_y fields are observed. Results for the xy-plane are the same as for the xz-plane, but with E_x and E_y swapped due to the different spatial distribution of the linear modes B_{1x} and B_{1y} (not shown).

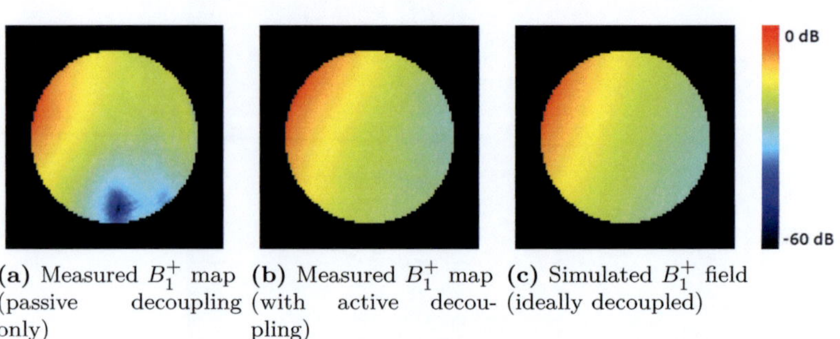

(a) Measured B_1^+ map (passive decoupling only)

(b) Measured B_1^+ map (with active decoupling)

(c) Simulated B_1^+ field (ideally decoupled)

Fig. 3.6: Measured and simulated B_1^+ fields of a single Tx element: The passive decoupling network only achieves partial decoupling (a) and the resulting map is slightly asymmetric. With active decoupling (b), the measurement is fully symmetric and corresponds well with the simulation (c).

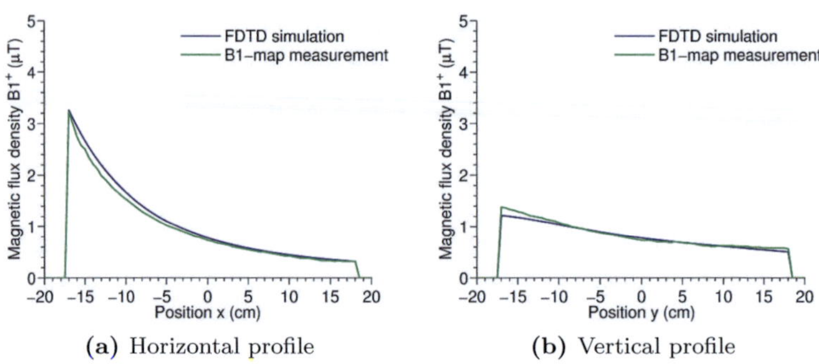

(a) Horizontal profile

(b) Vertical profile

Fig. 3.7: Measured and simulated B_1^+ profiles for a single Tx element with active decoupling: Measurement and simulation show good agreement.

3.2.2 Experimental validation of the E fields

For measurement of the electric fields, an MR-capable E field probe was built according to [70] and calibrated using a TEM cell. The probe consists of a fan-tailed dipole antenna with a diode detector and a high impedance connection to an oscilloscope to minimize coupling into the cable. Two sources of error were identified: 1) Varying the cable positioning and orientation inside the TEM cell led to measurement deviations up to 20%. 2) Ideally, the probe measures only the E field component which is in parallel to the antenna. However, during the calibration, the cross-sensitivity to orthogonal E fields was determined as up to 15%.

In the MR scanner, RF excitation was performed in quadrature mode using block pulses (duration 20 ms). The repetition time was set to TR $= 300$ ms to allow the probe to settle to zero between the pulses. For calibration of these pulses, the stimulated echo sequence from the systems power optimization was used with a small oil phantom in the isocenter. A unit PUC signal was found to correspond to $B_1^+ = 0.804\,\mu$T. The simulated RF fields were scaled to this amplitude for comparison.

The dominating electric field component E_z was measured along three different axes inside the bore of the empty scanner, as illustrated in Fig. 3.8. In each series, measurements were taken at intervals of 5 cm

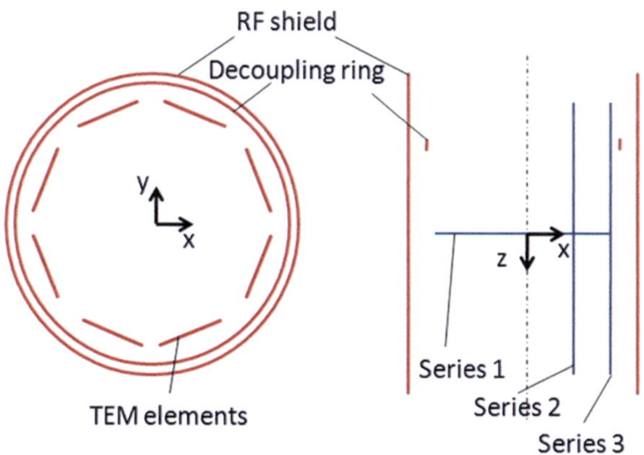

Fig. 3.8: Transverse (left) and axial (right) cross-sectional views of the MBC: The locations of the measurement series 1–3 in Figs. 3.9 to 3.11 are illustrated.

and repeated four times at each location to compensate for positioning inaccuracy.

In series 1, the E_z field was measured along the x-axis. As shown in Fig. 3.9, the measurement qualitatively and quantitatively agrees with the simulation. However, there is a slight asymmetry in the measured values, which is probably due to calibration errors of the individual TEM elements or due to imperfect RF decoupling. The maximum deviation of the mean to the simulation was $26.3\,\mathrm{V/m}$ at $x = -28\,\mathrm{cm}$, which corresponds to a deviation of 14.6% of the maximum E field in this series.

In series 2, the E_z field was measured along the z-axis at $x = 15\,\mathrm{cm}$. In this area, the E field is mostly determined by the spatial derivative of the B_1 field which changes rather smoothly. As shown in Fig. 3.10, the simulation generally conforms to the measurements, but overestimates the measured E field in the central region by up to 14.2%. The simulation results are shown with and without the decoupling ring, revealing only a marginal difference between the two cases along this axis.

In series 3, the measurement axis was shifted to $x = 28\,\mathrm{cm}$, which was the closest feasible distance to the bore wall. In this area, the E field is additionally constituted by the conservative E fields of the capacitors. The four series capacitors C_S occur as localized peaks in Fig. 3.11. The E field is not symmetric to $z = 0$ due to the decoupling network, which was not included in the initial simulation "FDTD without decoupling ring". The improved simulation with the decoupling ring matches the measurements to a maximum error of $29.0\,\mathrm{V/m}$ (14.9% of the maximum E field) for most locations. Only in the range $-25\,\mathrm{cm} \leq z \leq -15\,\mathrm{cm}$, the error increases up to $58.3\,\mathrm{V/m}$ at $z = -20\,\mathrm{cm}$ (29.9% of the maximum E field), even when the decoupling ring was included in the simulation. This is partly due to large E field gradient and the limited spatial resolution of the probe, but likely also due to limited accuracy of the simulation at this close proximity to the decoupling capacitor.

Overall, the simulated E field agreed well with the measurements. The error was not greater than 15% in the central area, where the E field is governed by the magnetically induced component. However, the simulation error may be significantly larger close to the capacitors, where the conservative E field depends strongly on the exact modeling of the capacitors.

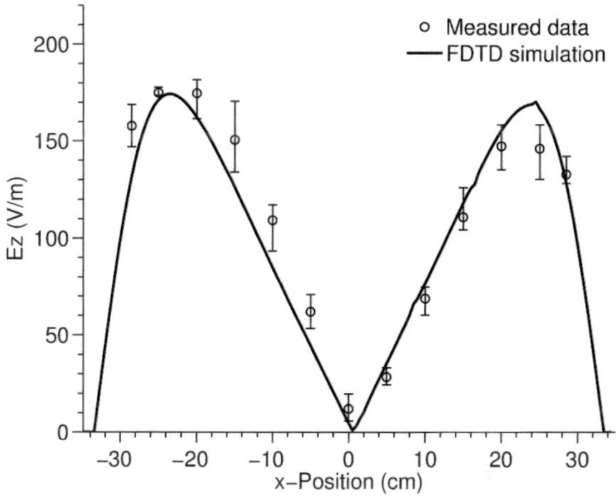

Fig. 3.9: Series 1: Measured E_z on the x-axis ($z = 0\,\text{cm}$, $y = -2\,\text{cm}$)
The errorbars indicate the minimum and the maximum value measured, as well as the mean (circle). Generally good agreement was observed, the measured E field distribution was however less symmetric than expected, which might be due to calibration errors. The simulation is also not perfectly symmetric, due to numerical errors.

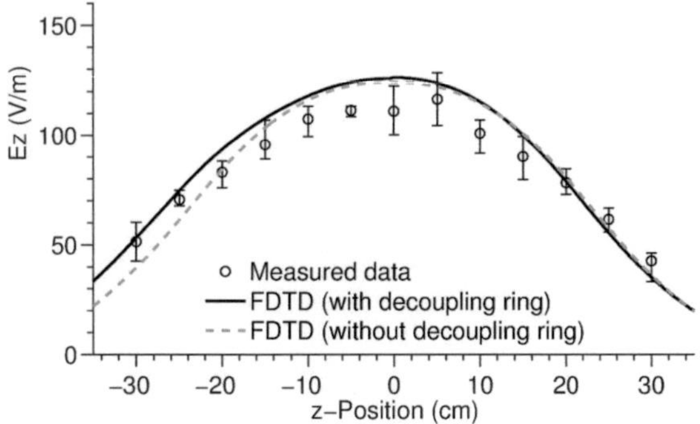

Fig. 3.10: Series 2: E_z in parallel to the z-axis ($x = 15\,\mathrm{cm}$, $y = -2\,\mathrm{cm}$)
The errorbars indicate the minimum and the maximum value measured, as well as the mean (circle). Good agreement was observed, independent of whether the decoupling ring is included in the simulation.

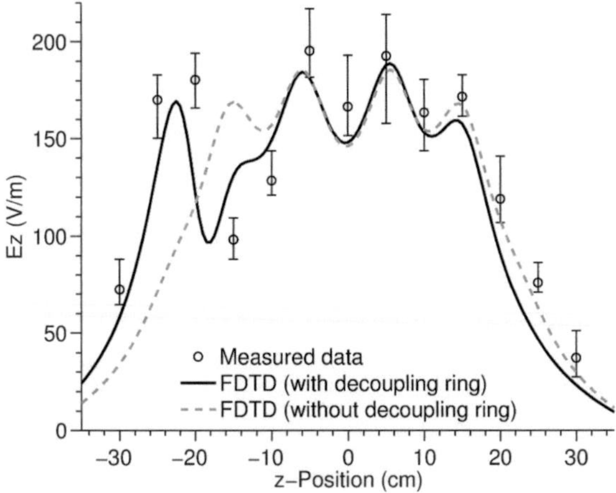

Fig. 3.11: Series 3: E_z in parallel to the z-axis ($x = 28\,\mathrm{cm}$, $y = -2\,\mathrm{cm}$)
The errorbars indicate the minimum and the maximum value measured, as well as the mean (circle). The location of the element capacitors C_{ele} is equally apparent in measurement and simulation. The steep local E field changes around the decoupling ring ($z = -20\,cm$) are only represented by the model that includes this feature.

3.2.3 Experimental validation of the global SAR

For validation of the global SAR, the simulated SAR values were compared to the resulting temperature increase measured in a phantom in the MR scanner. For this purpose, a phantom was first simulated in FDTD and then exposed to RF excitation in the MR scanner.

A spherical glass bottle with an inner diameter of $\varnothing 20\,\mathrm{cm}$ (i.e. a volume of 4.3 l) was used as the phantom and filled with a Na-Cl solution. The electrical conductivity was measured as $\sigma = 0.37\,\mathrm{S/m}$ at direct current and is frequency-independent for practical purposes [71] (error $< 1\%$ up to 128 MHz). The relative permittivity for this salinity can be calculated according to [71] as $\varepsilon_r = 78$. The phantom was insulated thermally using polystyrene plates and placed in the isocenter of the scanner. Then a fiber-optical temperature probe (Luxtron 790, LumaSense Technologies, Santa Clara, CA, USA) was placed in the center of the sphere and the phantom was allowed to settle thermally for several hours. Subsequently, an MR scan was performed for 31 minutes, using quadrature excitation at a predicted global SAR level of $2\,\mathrm{W/kg}$, as calculated using the Q-matrix from the FDTD simulation.

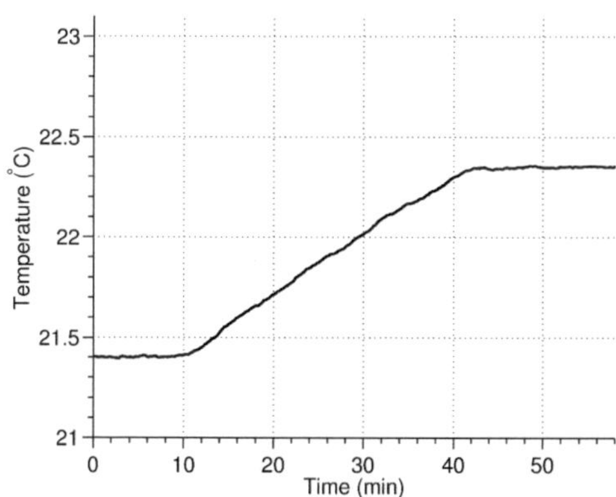

Fig. 3.12: Temperature rise in the calorimetry experiment. The MR scan was started at $t = 10\,\mathrm{min}$ and stopped at $t = 41\,\mathrm{min}$ and temperature was monitored with a fiber-optical probe. As expected, a linear temperature increase was observed.

The measured temperature response is shown in Fig. 3.12. Quantitatively, a temperature rise of $0.95\,\mathrm{K}$ was measured. From this temperature rise, the average SAR in the phantom can be estimated as $\mathrm{SAR} = c_H \frac{\Delta T}{\Delta t} = 2.13\,\mathrm{W/kg}$. This means that the measured SAR is 6.5% higher than the simulated SAR. Overall, the temperature increase due to the RF power delivered to the system agrees well with the simulation.

3.3 Conclusions

In this chapter, the generation of an FDTD model of the 8-channel Multix-Transmit Body Coil (MBC) was described. Numerical instabilities due to staircasing could be addressed by a closely symmetric design of the discretized strip-lines. A usual power-based calibration is not well suited for this load-sensitive Tx array. Instead, a novel calibration approach based on element currents and active decoupling was developed and validated. The simulated B_1 field, E field, and temperature increase were consistent with the measurements.

Chapter 4

Patient-specific SAR models

In recent years, several generic human body models have been developed. Among these, the Visible Human Male and Female [72–74], the Virtual Family [75], and the NORMAN and NAOMI models [76, 77] are most prevalent. Despite of the variety of these models, the SAR prediction accuracy is inherently limited since these models do not match the individual patient and have never been validated. The patient-dependency of the local SAR leads to a systematic error of any EM simulation using generic body models, as has been pointed out by several researchers. These errors arise from factors such as body size and shape [78–80], gender and body-fat distribution [74] as well as body posture and position within the MR scanner [81].

It is generally reasonable to use multiple human body models for safety assessment to account for these patient-dependent variations [6]. Anatomically poseable versions of the Visible Human Male and Female (Varipose, Remcom Inc., State College, PA, USA) as well as the Virtual Family models (SEMCAD X Poser Package, SPEAG, Zurich, Switzerland) have been developed for this purpose. Such skeleton-based modifications, however, do not represent different body shapes or varying fat and muscle content within the body.

Furthermore, there is currently no consensus about the required model complexity. It is self-evident that a high number of segmented body tissues and a maximal spatial resolution is desirable to achieve high accuracy. Currently, the highest model detail is available for the Virtual Family, offering up to 0.5 mm spatial resolution and about 80 different tissue types. From a practical perspective, a lower spatial resolution is desirable due to memory and processing constraints. For example, the

---

Based upon: H. Homann, P. Börnert, H. Eggers, K. Nehrke, O. Dössel, and I. Graesslin, "Toward Individualized SAR Models and In Vivo Validation," *Magn. Reson. Med.*, vol. 66, pp. 1767–1776, 2011.

computation time of FDTD scales with the 4th power of the spatial resolution. Especially for parallel transmit MR systems, where simulations have to be carried out repeatedly for several Tx coil elements, this is an important factor.

In this chapter, first the required model detail is investigated with respect to spatial resolution and the number of body tissue classes. Then, the hypothesis is proposed that the local SAR values can be sufficiently represented by simplified models which distinguish only between fat, water-rich, and lung tissues. Supported by this, a novel process for generation of individualized body models for SAR calculations is proposed. Finally, a first *in vivo* validation of EM simulations is presented. For all simulations, the MBC was driven in quadrature mode. All simulated Tx fields were scaled to a B_1 field of $1\,\mu\text{T}$, averaged over the body model in the transverse isocenter slice.

4.1 Numerical resolution

For large homogeneous objects, a numerical grid resolution of $\Delta x \leq \lambda_{\min}/10$ (where $\lambda_{\min}$ is the shortest wavelength) is generally considered as sufficiently accurate for modeling the RF wave propagation on a numerical grid [57]. In addition, the grid-spacing should also be much smaller than the skin-effect depth to accurately model the current distribution within lossy materials. At $128\,\text{MHz}$, these values are $\lambda_{\min} \approx 300\,\text{mm}$ and $\delta_{\text{skin}} \approx 50\,\text{mm}$ in muscle tissue, indicating that a grid resolution of $5\,\text{mm}$ might be sufficient for EM simulations.

However, for simulations of the human body, the heterogeneous anatomy needs to be modeled accurately enough to represent the current paths through the body, especially in the proximity of potential local SAR hotspots. For the large organs, a rather coarse grid may be acceptable, whereas different requirements may apply in more delicate regions, especially in the head [82]. An isotropic grid resolution of $5\,\text{mm}$ has been reported to be suitable for RF simulations at $64\,\text{MHz}$ when using an MR saddle head coil, loaded with the Visible Human Male [83].

Methods

The question of a reasonable grid resolution for SAR simulations of a $3\,\text{T}$ body coil is addressed in this section. The Visible Human Male model was rescaled from a 1 mm voxel dataset to isotropic grid resolutions of 2.5 mm, 3.5 mm, and 5 mm. The resulting body models were simulated inside the model of the MBC for an abdominal scan position. The mesh sizes were 300×300×770 cells at 2.5 mm grid resolution, 230×230×566 cells at 3.5 mm, and 170×170×400 cells at 5.0 mm. The dielectric tissue properties at 128 MHz were calculated according to Gabriel *et al.* [84] as listed in Appendix A.5. The tissue density was uniformly set to $1\,\mathrm{g/cm}^3$. The model of the RF body coil was initially created and tuned at a grid resolution of 5.0 mm. This geometric coil model was re-meshed to the finer mesh resolutions. This procedure was verified to have no effect on the coil's resonance frequency. All results were normalized to $1\,\mu\mathrm{T}$, averaged over the body in the isocenter slice.

Results

The local SAR distribution for the Visible Human Male at the different grid resolutions and tissue segmentations is compared in Fig. 4.1 and the relevant numerical SAR values are listed in Tab. 4.1. The general pattern of the local SAR is highly similar for all three spatial resolutions. The global SAR values are almost constant, independent of the grid resolution. The maximum local SAR for a single mesh cell increases with resolution. When averaging over 1 g or 10 g of tissue, this trend is no longer observed. When averaging over 10 g, the numerical error due to grid resolution is in the range of 20%. Independent of the grid resolution, the maximum local SAR occurred at the same location in the muscle tissue of the left arm, not varying by more than 5 mm independent of the resolution. Using the GPU cards, computation times for each Tx coil element were 45 min at 5.0 mm spatial resolution and 2 hours at 3.5 mm resolution. The 2.5 mm grid exceeded the memory size of the GPUs and had to be calculated on the CPU instead, resulting in a computation time of 104 hours per Tx coil element. The SAR calculation and averaging over 10 g was performed in less than one minute.

Discussion

The variation of the grid resolution did not lead to considerable changes in the body-averaged SAR. The general local SAR pattern was highly similar for all resolutions investigated, but the exact numerical values varied even when using an averaging mass of 10 g. Even though the

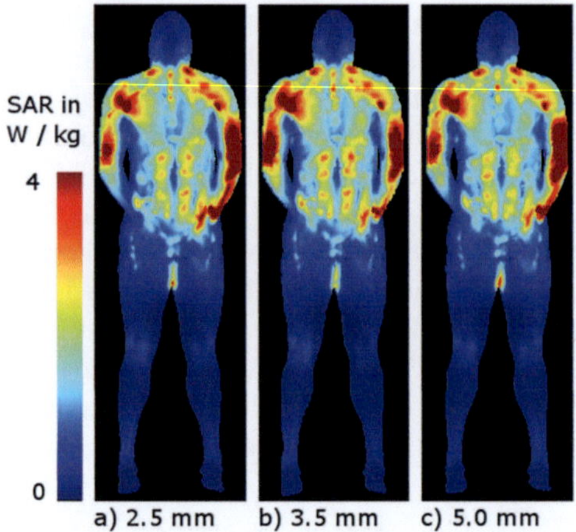

Fig. 4.1: SAR distributions for the Visible Human Male at different grid resolutions: Shown are maximum intensity projections, averaged over 10 g of tissue. The local SAR pattern shows little variation when reducing the grid resolution from 2.5 mm to 5.0 mm.

distance between the arms of the Visible Human Male and the RF body coil was relatively small (2.5 cm at the closest point), the local SAR in the arms showed little variation with the spatial resolution. In contrast, a different positioning of the lower arms leads to strong variations of the local SAR [85].

Overall, the error due to grid resolution is relatively small compared to patient-dependent variations of the local SAR values. This suggests that a 5.0 mm grid resolution is an acceptable compromise for SAR modeling of RF body coils at 3 T.

Tab. 4.1: SAR values (in W/kg) for the Visible Human Male at different grid resolutions, normalized to $B_{1,\mathrm{rms}} = 1\,\mu\mathrm{T}$ at quadrature excitation.

Grid resolution	whole-body SAR	unaveraged local SAR	max. 1 g local SAR	max. 10 g local SAR
2.5 mm	0.54	35.68	13.76	8.88
3.5 mm	0.57	24.91	11.63	6.98
5.0 mm	0.58	22.78	11.90	8.79

4.2 Soft-tissue representation

This section addresses the required detail in the soft-tissue segmentation for accurate modeling of the dielectric tissue properties of a human body. Overall, even a large uncertainty (up to a factor of 2) in permittivity and conductivity was reported to have little influence on the whole-body SAR estimation, whereas no direct correlation can be predicted for the local SAR [86]. More recently, Gabriel and Peyman [87] presented an error analysis that attributes the large differences in published results to random uncertainties and quantified the error in their dielectric values as less than 10% standard deviation of the mean.

While several highly detailed human body models exist, the question which tissues in the segmentation are essential for accurate SAR estimation has rarely been addressed. Comparing the dielectric properties of human body tissues reported by Gabriel *et al.* [84] as visualized in Fig. 4.2, most organs of the human body have dielectric properties similar to muscle tissue ($\sigma = 0.72\,\mathrm{S/m}$, $\varepsilon_r = 63.5$ at $128\,\mathrm{MHz}$) within an error of 40%. Fat tissue is distinguished in that its conductivity and permittivity are comparably small, such that this tissue hardly influences the RF wave propagation. The lungs and bones exhibit dielectric properties in the intermediate range between muscle and fat. Some ionic body fluids, such as the cerebrospinal fluid (CSF), show a significantly higher conductivity than the body's organs. Quantitatively, fat and muscle contribute most to the total body mass, but also show the highest variation among patients [88], whereas the remaining tissues occur in much smaller fractions. Consequently, van den Bergen *et al.* [89] reported no significant differences in the SAR values in a human torso model when replacing all tissue types other than bone, fat, and lung by muscle, facilitating automated segmentation by thresholding from CT scans.

During RF exposure, power deposition occurs rather in highly-conductive (muscle, liver, CSF, etc.) than in low-conductive (fat and bone) tissues [90]. This effect is well known in medical diathermy, where RF-induced eddy currents are commonly used to achieve selective heating of conductive tissues, e.g. for muscle relaxation. The eddy current pathways through the body are composed of conductive tissues and the available cross-section along a current path is determined by the surrounding low-conductive tissue, i.e. the body fat distribution [18]. The above relations lead to the hypothesis that the proportions and the distribution

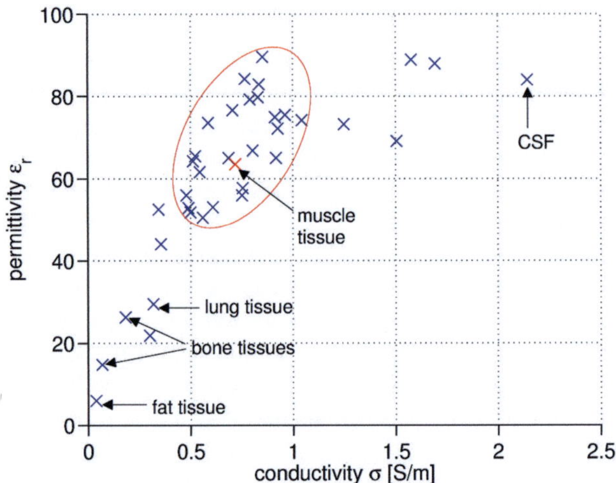

Fig. 4.2: Dielectric tissue properties at 128 MHz according to Gabriel *et al.* [84], every cross corresponds to one tissue class. The properties of most organs are similar to muscle tissue. Fat tissue shows little interaction with the RF field.

of body fat and muscle tissue have a distinguished impact on the wave propagation and eddy current pathways in the human body, whereas other organs can potentially be treated as similar to muscle tissue for EM simulations.

Methods

To test the hypothesis whether muscle tissue is a reasonable representation of all water-rich body tissues, three different simulations were carried out for the Visible Human Male and Female: a) using the original models (number of tissue types: 36 in the male model, 33 in the female model), b) with the dielectric properties of fat assigned to bones and the properties of muscle assigned to all other tissues except the lungs (in the following referred to as "muscle-fat-lung model"), and c) replacing the fat segments by muscle as well (referred to as "muscle-lung model"). An isotropic grid resolution of $5\,\mathrm{mm}$ was used and again local and whole-body SAR values were calculated. Again, all results were normalized to $1\,\mathrm{\mu T}$.

Results

A comparison of the original Visible Human Male and Female models with the muscle-fat-lung models is shown in Fig. 4.3. The local SAR patterns of the muscle-fat-lung models can hardly be distinguished from the original. The hotspots occur at the same locations, varying only marginally in shape. Quantitative results are listed in Tab. 4.2. Without averaging, the local SAR values differ significantly between the models. However, when averaging over $10\,\mathrm{g}$ of tissue, the differences in local SAR are clearly reduced (7.5% and 2.6% at the maximum hotspot for the male and the female model, respectively). In contrast, the hotspots in the muscle-lung models appear blurred or are not present at all (differences $> 100\%$).

Discussion

A close agreement of the original Visible Human models with the muscle-fat-lung models was observed. This indicates that not all detail provided by present body models is required for SAR calculations and that the number of tissues can be significantly reduced. Neglecting the distinction

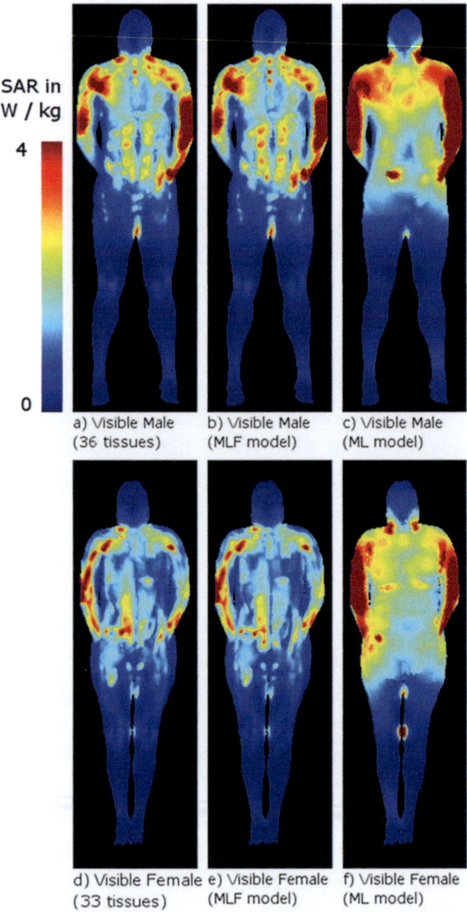

Fig. 4.3: SAR distributions for the Visible Human Male and Female with different soft-tissue segmentations: Shown are maximum intensity projections, averaged over 10 g of tissue. For both models, the local SAR pattern shows little variation when reducing the number of tissue classes to the muscle-fat-lung (MFL model). Not differentiating between muscle and fat in the muscle-lung (ML) model does not correctly represent the local SAR hotspots.

between water-rich and fatty tissues in the muscle-lung model, however, led to an unacceptable blurring of the SAR hotspots. This blurring can be explained by the fact that the eddy current paths through the human body are determined by the distribution of conductive tissues. At locations where the conductive paths become narrow between regions of low-conductive (fatty) tissues, hotspots can occur [18]. This is not represented in the simplistic muscle-lung models which do not describe the eddy current pathways and hence cannot correctly resolve the local SAR patterns. Due to this blurring of the hotspots, lower overall SAR levels might be expected. However, the SAR in the muscle-lung models is higher than in the original model. This is due to the increased dielectric loading from the highly conductive model and an increased RF input power when normalizing the results to $B_1^+ = 1\mu\mathrm{T}$.

Furthermore, the validity of muscle-fat-lung models is also supported by the *in vivo* study in the following sections, where promising agreement was obtained between the simulated RF fields based on the volunteer models and the measured B_1 maps. However, the simplification down to just three tissue classes might be too drastic for accurate representation of all hotspots in the human body. Especially, the representation of tissues with dielectric properties significantly different than muscle, such as CSF, requires further research.

Tab. 4.2: SAR values (in W/kg) for the Visible Human Male and Female models with different soft-tissue segmentations, normalized to $B_{1,\mathrm{rms}} = 1\,\mu\mathrm{T}$ at quadrature excitation.

Model	whole-body SAR	unaveraged local SAR	max. 1 g local SAR	max. 10 g local SAR
Visible Human Male:				
Full segmentation	0.58	22.78	11.90	8.79
Only muscle, fat, lung	0.59	32.14	12.96	8.13
Only muscle, lung	0.87	49.09	41.85	29.67
Visible Human Female:				
Full segmentation	0.51	11.44	7.04	4.70
Only muscle, fat, lung	0.51	18.76	9.12	4.82
Only muscle, lung	0.92	22.84	16.43	12.67

4.3 Body model generation based on water-fat separated MR data

Methods

A process for generation of individualized body models is proposed in this section, comprising the following steps: 1) The acquisition of a chemical-shift encoded whole-body MR dataset, 2) a voxel classification based on water-fat separation, and 3) the generation of an FDTD model.

The exact resonance frequency of the hydrogen nuclei depends on their chemical binding. Water and fat exhibit a chemical shift of 146 Hz/T. In each voxel containing water and fat, the resulting signal varies in amplitude and phase, depending on the contributing water and fat fractions and the echo time. If two or more MR images with different echo times are acquired, the contributing water and fat fractions can be reconstructed using vector geometry [91]. Such approaches are referred to as Dixon methods. In contrast to chemical-shift selective approaches, Dixon methods allow for simultaneous acquisition of water and fat images and are more robust against inhomogeneities of the main field B_0 to achieve optimal results over a large field of view.

In this work, whole-body 3D imaging was performed at the 3 T scanner with the MBC (5 volunteers) and at a conventional 1.5 T scanner (9 volunteers). A multi-gradient echo sequence was applied in a conventional multi-station scanning approach employing the body coil for reception. Three echoes (echo time intervals ΔTE $= 1.5$ ms and 0.7 ms at 1.5 T and 3 T, respectively) were sampled after each RF excitation pulse, using a bipolar readout gradient to increase scan efficiency [92]. The pixel bandwidth was about 10 times the water-fat shift, so that chemical shift displacement effects were negligible. An isotropic grid resolution of 5 mm was chosen; a flip angle of 10° and a repetition time of TR $= 5.6$ ms were used. The field of view at each station was set to $520 \times 390 \times 160$ mm^3. For validation, B_1 field maps were acquired from the volunteers scanned in the 3 T MBC directly after the whole-body scan, as described later in Section 4.4.

The 3D data were reconstructed for each station and each echo individually. In this work, a three-point Dixon approach was used, similar to the one described in [93]. To compensate for phase errors resulting from the bipolar readout gradient, an eddy current correction was applied [94]. The water-fat-separated images of the different stations were finally combined to form whole-body water and fat images.

The voxel classification was performed as follows: Each voxel was labeled either as "water", if its intensity in the water image was greater than the intensity in the fat image, or as "fat" in the opposite case. For segmentation of the lungs and the background, a mixture model [95] was applied to find an intensity threshold. Two Gaussian distributions were used to model the water- and fat-dominated tissues. An exponential distribution was chosen to model the background noise. The model parameters of these three distributions were optimized iteratively by an Expectation-Maximization algorithm, as described in detail in Appendix A.3. An example of the resulting distributions is shown in Fig. 4.4.

All voxels with an intensity below the threshold were joined to groups of connected voxels. The largest group was then labeled as "background", the second and third largest groups were labeled as the two "lungs". Usually, some more groups remained due to areas of low signal intensity inside the human body, e.g. in cortical bone or in air filled cavities (stomach, intestines, or paranasal sinuses). These regions are typically low-conductive and were also labeled as "fat".

Results

Whole-body imaging took about 5 minutes and was mostly limited by repeated acceleration and deceleration of the patient table. Consistent 3D image quality and robust water-fat separation were observed for almost the entire body. Visual inspection of the models and the original images showed that the classification into water and fat agreed with the anatomy. The brain and the abdominal organs were represented by muscle tissue and the bones were mostly modeled by fat. However, structures of a size similar to or smaller than the scan resolution were not accurately represented. In such cases, the surrounding tissue determines the classification due to the partial volume effect. This can be observed in Fig. 4.5, where the skull as well as the tibia and fibula are partially classified as muscle tissue. Some minor segmentation errors are also observed as reduced intensity and swapping artifacts in some of the volunteers in the outer regions of the arms (cf. Fig. 4.5). These areas were located at the borders of the scanner's imaging volume, where the main field homogeneity locally changes very rapidly, making water-fat separation difficult. This effect was less pronounced at 1.5 T than at 3 T due to better main field homogeneity.

The background threshold was reliably found by the Expectation-Maximization algorithm within a few iterations and was robust against variations of the initial distribution parameters. Visual inspection further showed that the threshold allowed a valid detection of the background for all volunteers. The same threshold also identified the lungs correctly.

Discussion

The presented approach for generating human body models based on 3D water-fat-resolved MR data reflects the importance to differentiate between highly-conductive water-rich and low-conductive fatty tissues as well as the lungs for accurate representation of the eddy current paths in the human body. The presented approach is relatively robust and suitable for future automation of the model generation process. This represents a major advantage over traditional, mostly manual segmentations.

Modelling of smaller structures in the range of the voxel size was found to be incomplete, which will lead to inaccurate representation of the current paths, e.g. in the head due to inaccurate segmentation of the skull. To avoid such errors from partial volume effects, a higher resolution might be required in the concerned areas. Especially for SAR simulations of the head, further studies are required. Furthermore, the residual water-fat swapping artifacts at the arms remain to be addressed for automated modeling of bigger patients. In the future, this problem might be overcome by the advancement of wide-bore and better magnets as well as dedicated water-fat separation algorithms.

Scan efficiency of the approach might be improved further by the use of moving table approaches [96], advanced two-point-Dixon methods [94], or compressed sensing [97]. Possibly, further improvement could be obtained by a more elaborated segmentation in temperature critical regions, such as the eyes or the testes.

One application area for such models is RF coil design, where the dielectric loading by the patient significantly affects tuning and power matching. Furthermore, SAR optimization during RF coil design is gaining importance with the advance of higher field strengths due to intrinsically higher SAR. As the proposed three-compartment models consist

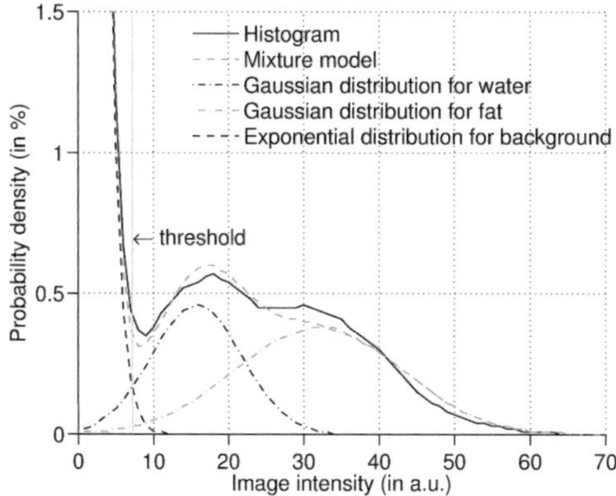

Fig. 4.4: Generation of the muscle-fat-lung models: To find a threshold that identifies the background and the lungs, a mixture model is fit to the histogram by an Expectation-Maximization algorithm. The threshold is defined as the intensity where the distribution functions of water and of the background cross each other.

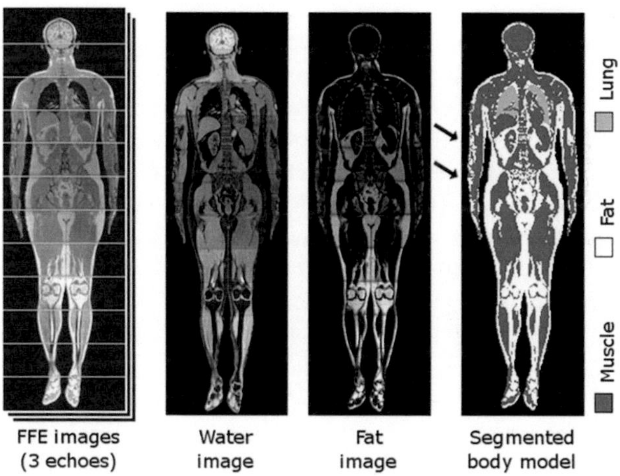

Fig. 4.5: Generation of the muscle-fat-lung models: From the water-fat-separated Dixon images, an intensity-based voxel-by-voxel classification into pure water and fat voxels is achieved. Some smaller artifacts in the water-fat separation at the arms are indicated by arrows.

of relatively large homogeneous regions. They are hence well suited for a tetrahedral representation with few elements. With such models, the finite elements method (FEM) or the method-of-moments (MoM) might become suitable alternatives to FDTD simulations that are not restricted to a staircased geometry and are hence better suited to model the fine oblique structures of RF coils.

4.4 *In vivo* validation

The voxel models generated at the 3 T scanner were imported into the FDTD simulation software and carefully placed inside the model of the MBC. The model position was adjusted to match the volunteer's body orientation and position during the acquisition of the B_1 field maps. Dielectric properties according to [84] were assigned to the segments as listed in Appendix A.5. Subsequently, FDTD simulations were performed for SAR estimation and experimental validation.

For validation of the EM simulations by means of MR, B_1 mapping was performed. A similar procedure was applied earlier in a human corpse, using a multi-flip angle method and an additional T1 mapping scan for error correction [98]. However, due to relatively large repetition times, that approach is not feasible *in vivo*. In this study, the actual flip-angle imaging technique (AFI) was used for B_1 mapping, which is based on a spoiled steady-state protocol with dual alternating TR and is inherently T1-insensitive for $TR_1, TR_2 < T1$ [11, 67]. The B_1 field maps were acquired directly after the whole-body scan in the identical posture of the volunteer within a single breathhold. The protocol parameters were $TR_1/TR_2 = 20/100$ ms, $TE = 2.3$ ms, and flip angle $60°$ at a scan resolution of $5 \times 5 \times 15$ mm^3.

The simulated and the measured B_1 field maps are compared in Fig. 4.6. Fig. 4.6a shows a validation for different anatomical regions in a selected volunteer. The general field pattern in the simulation agrees well with the measurements, indicating that the RF wave propagation is reasonably represented by the muscle-fat-lung model. This can be seen in the legs, where muscle and fat tissue clearly dominate, as well as in the abdomen, where the intestines have similar dielectric properties as muscle tissue. Also in the head, where this would not necessarily be expected, a good agreement was found. An explanation for this is that the brain is entirely represented as "muscle" in the models, which has

54

similar dielectric properties as gray matter, the major component of the brain.

Fig. 4.6b shows changes of the magnetic field for another volunteer when positioning arms either along the body or on top of the belly. The general field pattern in the torso is similar for both arm positions, but some differences can be seen at the sides of the torso, where B_1 is higher with the arms stretched out along the body. The standard deviations of the difference images ranged from 14.0% to 22.6%.

Overall, the comparison of the simulated and the measured B_1 fields showed quantitative and qualitative correspondence which presents, to our knowledge, the first *in vivo* validation of RF field simulations in the human body. The standard deviation of the error ranged up to about 20%, however locally higher deviations were observed. Some of these deviations might be attributed to errors in B_1 mapping or to motion of the volunteers between the scans.

4.5   Conclusions

In this chapter, a novel approach to the generation of body models for SAR simulations was presented. This approach considers that eddy current paths inside the patient body are mostly determined by the water-fat distribution. This suggests that the number of tissue segments can be significantly reduced. Body models with just three tissue classes ("fat-like" tissue – including bone, "muscle-like" tissue, and the lungs) at a spatial resolution of 5 mm showed consistent results for an RF body coil at 3 T.

An MR procedure for acquiring such models based on water-fat-separation was presented, which is robust enough for future automation. A validation of the simulation results was performed by comparison with B_1 field maps, showing quantitative and qualitative agreement of the RF field pattern for several volunteers in different body positions.

Overall, a more realistic representation of the actual SAR for a specific volunteer can be expected from these individualized body models than provided by existing generic body models. In a clinical setting, generating specific body models for every patient is not practical today due to simulation times in the order of a few hours. For this reason, the model generation and simulation process needs to be separated from

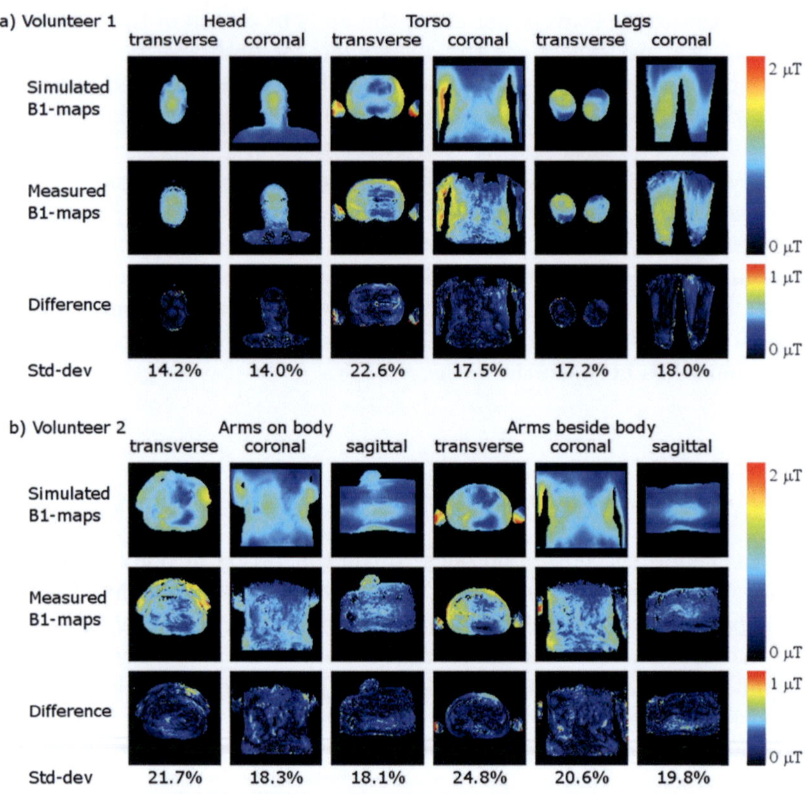

Fig. 4.6: Comparison of simulated and measured B_1 fields: a) B_1 maps at different patient positions in the scanner show a similar pattern. The standard deviation of the difference is given below the respective images. b) Different placements of the arms result in small variations of the B_1 field, especially at the sides of the torso. All images were normalized to $B_1 = 1\,\mu\text{T}$, averaged over the transverse slice.

the actual MR examination. Instead of creating patient-specific models, the presented approach can be used to create a database representing a broad variety of patients and covering different anatomies and body poses. This allows to select the most suitable model for each individual patient, depending on his/her height, weight, gender, body positioning within the RF coil, and other parameters for appropriate SAR estimation.

The presented approach supports the rapid generation of body models for patient-specific and more accurate SAR simulations. Even though online-modeling from a patient-specific pre-scan seems not practical today, similar approaches are current clinical practice for attenuation correction in PET-MR [99]. In future, accurate SAR prediction will certainly gain importance, especially with the advance of ultra-high-field MRI ($\geq 7\,$T). Hence, patient-specific generation of SAR models may become a realistic option and advanced numerical methods [100] can facilitate simulations almost in real-time.

Chapter 5

Local SAR management

So far, this work has investigated the generation of accurate models of a multi-channel RF coil and of patients as well as SAR calculations based on such models. The following two chapters will discuss how these models can be employed to minimize the SAR by a suitable design of the RF sequence.

The ability to control the RF field spatially by parallel transmission can not only be applied to optimize the RF field homogeneity but also has the potential to reduce the SAR simultaneously [101,102]. This can be beneficial in MR sequence design, e.g. to increase the allowed range of flip-angles and repetition times for improved contrast and SNR or to reduce the total scan time. In this work, an approach for optimizing the RF sequence such that the SAR limits are met is referred to as *SAR management*. In principle, such SAR management can employ the same SAR model as used in the scanner for SAR calculation.

For reliable SAR calculation and SAR management, the choice of an appropriate human body model is an important issue. For MR systems with a single transmit coil, significant variations of the SAR values have been reported when comparing different models [74,78,79,103] and different positioning within the MR scanner [39,81]. However, these aspects have so far rarely been discussed in the context of parallel transmission.

In Section 5.1, the potential of SAR management for RF shimming is evaluated for the 8-channel MBC based on FDTD simulations. Sec-

Based upon: H. Homann, I. Graesslin, H. Eggers, K. Nehrke, P. Vernickel, U. Katscher, O. Dössel, and P. Börnert, "Local SAR Management by RF Shimming: A Simulation Study with Multiple Human Body Models," *Magn. Reson. Mater. Phy.*, (accepted), 2011.
and H. Homann, P. Börnert, O. Dössel, and I. Graesslin, "A Robust Concept for Real-Time SAR Calculation in Parallel Transmission," in *Proceedings of the ISMRM, Montreal, Canada*, p. 3843, 2011.

tion 5.2 describes a robust approach that allows incorporating multiple models with only a minor increase of computational effort.

5.1 Local SAR management for RF shimming

To investigate the patient-dependency of the SAR, simulations were performed for nine human body models, representing a relatively broad range of male adults. These models were generated from water-fat separated whole-body MR images as described in the previous chapter.

The approach of the section is twofold: In a first step, the effect of RF shimming on the SAR is studied without imposing any constraints. The influence of the region of interest (ROI) chosen for RF shimming is also addressed. In a second step, model-based SAR constraints are included using a novel optimization approach. This study limits itself to a discussion of the local SAR, since the maximum local SAR often represents a more restrictive condition than the global SAR [36], which can be estimated from power measurements [32].

5.1.1 Theory

Before applying an RF pulse *in vivo*, the pulse sequence has to be validated to conform with the SAR limits. To achieve this, the SAR is estimated using an appropriate model. In conventional single-transmit systems, a violation of the SAR limits would require a reduction of the average RF power, e.g. by prolonging the RF pulse, increasing the repetition time TR, or reducing the flip-angle.

Parallel transmission offers a new degree of freedom for SAR management by spatial control of the RF field, to avoid such changes in the sequence design. Recently, a number of local SAR management techniques have been proposed. These can be classified as regularization-based approaches [104–106] and approaches that enforce inequality constraints [101, 107, 108].

The regularization-based approaches use multiple regularization parameters, either per channel [104] or for selected cells of the simulation mesh [105, 106], to get a handle on the local SAR distribution. This can, however, become computationally demanding as numerous regularization parameters need to be optimized iteratively. This is mostly addressed by updating the regularization parameters iteratively until the SAR constraints are met. However, such a heuristic approach does not guarantee to give the optimal B_1 distribution.

Optimization problems with strict inequality constraints can be efficiently addressed by interior-point methods, such as the Lagrangian-dual method or the barrier method [109]. These allow reducing the constrained problem to a sequence of unconstrained problems. When constraints on the local SAR are considered, the number of inequality constraints becomes very large. Solving the problem with multiple SAR constraints via the Lagrange-dual problem drastically increases the number of unknowns, hampering the iterative solution. Instead, a logarithmic barrier method is proposed in this work, offering a numerical effort which increases only linearly with the number of local SAR constraints.

The optimization problem for RF shimming is given by:

$$||\,|\mathbf{Aw}| - \hat{\mathbf{b}}||_2^2 \quad \rightarrow \quad \min \tag{5.1}$$

$$\text{s.t. } \mathbf{w}^H \mathbf{Q}_i \mathbf{w} \cdot a_{\text{rms}}^2 - \text{SAR}_{\text{max}} \quad \leq \quad 0 \qquad \forall i \tag{5.2}$$

In the above equations, $\mathbf{A}$ contains the magnetic transmit field sensitivities $\mathbf{S}_{B1}(\mathbf{r})$, stacked for all voxels in the region of interest (ROI); $\mathbf{b}$ is the desired B_1^+ field. The circumflex denotes that only the amplitude is used as the target because the phase of the RF transmit field is of little interest in many MR applications. This helps to improve RF performance even further. The side condition Eq. 5.2 is introduced to enforce regulatory SAR limits. Each inequality constrains the local SAR at one voxel i in the compressed model.

The objective function Eq. 5.1 is non-linear, but the magnitude least-squares (MLS) algorithm [110] can be applied to solve this problem by sequentially optimizing the phase of $\mathbf{b}$ using the local phase-exchange method [111]. At each iteration, this requires solving the following quadratically-constrained quadratic-program (QCQP) for its optimizer $\mathbf{w}$:

$$||\mathbf{Aw} - \mathbf{b}||_2^2 \quad \rightarrow \quad \min \tag{5.3}$$

$$\text{s.t. } \mathbf{w}^H \mathbf{Q}_i \mathbf{w} \cdot a_{\text{rms}}^2 - \text{SAR}_{\text{max}} \quad \leq \quad 0 \qquad \forall i \tag{5.4}$$

As $\mathbf{A}^H \mathbf{A}$ and the Q-matrices are positive definite, the problem is convex and hence is a second-order cone program (SOCP) (cf. [108]). A logarithmic barrier method was chosen to incorporate the inequality constraints and the objective function into a single cost function $C(\mathbf{w})$ as:

$$C(\mathbf{w}) = ||\mathbf{Aw} - \mathbf{b}||_2^2 - \frac{1}{t} \sum_i \ln(\mathrm{SAR_{max}}/a_{\mathrm{rms}}^2 - \mathbf{w}^H \mathbf{Q}_i \mathbf{w}) \qquad (5.5)$$

The logarithmic barrier causes the cost to increase to infinity as the SAR at any mesh cell i approaches its limit. The steepness of the barrier is controlled by the parameter t. In this work, t was iteratively increased from 0.1 to 1000 in steps of power of 10, where the final parameter $t = 1000$ represents maximum steepness. An extensive discussion of this approach is given in [109]. For RF shimming, the number of unknowns is small and Newton's method lends itself for optimization with quadratic convergence and was hence applied in this study. The algorithm and the required derivatives of the cost function are provided in Appendix A.4.

5.1.2 Experiments and simulations

Numerical simulations

To evaluate the effect of RF shimming on the SAR in different individuals, a number of dielectric body models is required. Nine volunteer models were simulated inside the MBC model in a comparable scan position with the kidneys in the isocenter, as illustrated in Fig. 5.1. The volunteers represent a relatively broad range in terms of body size, weight, and body fat content, as listed in Tab. 5.1.

FDTD simulations were carried out for each transmit element individually and the simulation results were normalized and post-processed as described in Chapter 2. Q-matrices were generated using an averaging mass of $10\,\mathrm{g}$. The matrices were then compressed using an uncertainty term ε of 20% of the maximum local SAR at quadrature excitation. The compressed Q-matrices are then stored in a database on the host computer of the MR scanner.

Evaluation of different RF shimming strategies

In this study, RF shimming was performed based on the magnetic field sensitivities $\mathbf{S}_{B1}(r)$ obtained from the FDTD simulations. If not stated otherwise (Fig. 5.5), the simulations were performed for an RF pulse sequence with a root-mean-square (rms) value of $B_{1,\mathrm{rms}} = 2\,\mu\mathrm{T}$ and SAR

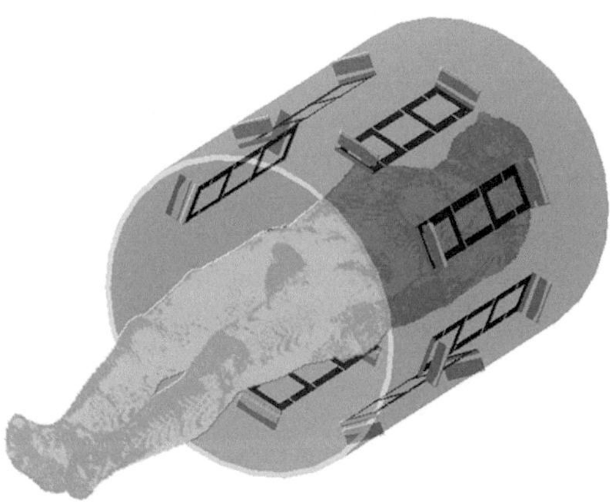

Fig. 5.1: Body model placed inside the model of the multi-channel body coil (MBC) for FDTD simulation: All body models were positioned at an abdominal scan position.

Model	Height / m	Weight / kg	Body fat content
Volunteer 1	1.73	65	27 %
Volunteer 2	1.82	68	22 %
Volunteer 3	1.92	70	30 %
Volunteer 4	1.91	71	26 %
Volunteer 5	1.82	75	30 %
Volunteer 6	1.80	86	43 %
Volunteer 7	1.92	93	39 %
Volunteer 8	1.85	96	33 %
Volunteer 9	1.88	103	54 %

Tab. 5.1: Overview of relevant volunteer data: The body fat content was calculated from the segmented water-fat separated models as the fraction of fat voxels (including bone) per total number of voxels. Volunteers were ordered by weight, the same order is used for the following figures.

values were calculated for the uncompressed (original) SAR model. The following RF shimming cases were considered and compared to quadrature excitation:

RF shimming without constraints: First, RF shimming was performed without any SAR constraints or regularization. The transverse isocenter slice was used as a region of interest (ROI) for B_1 homogeneity optimization. Two different cases were investigated: a) the cross-section of the arms and the torso were included in the ROI and b) only the torso was included.

RF shimming with SAR constraints: Then, constraints on the local SAR were included in the RF shimming algorithm as described in the previous section. The SAR limits were set to 10 and 20 W/kg for the torso and the extremities, respectively, and only the torso was included in the ROI. Again, two different cases were investigated: a) The optimization was performed for each of the volunteer models separately, using the individualized SAR models. b) Another optimization was performed using the biggest of the volunteer models as a "generic" model for SAR-constrained RF shimming of all other volunteers. This latter case was performed to obtain a more realistic scenario, as the use of individualized models is currently not practical in a clinical setting.

5.1.3   Results

Model compression

The number of mesh cells in the body models ranged from 310,000 to 540,000, increasing with body weight. The model compression algorithm with an uncertainty term of $\varepsilon = 20\%$ reduced this down to 80 to 714 representative cells.

SAR at quadrature excitation

The local SAR distribution for all volunteers at quadrature excitation is compared in Fig. 5.2a as maximum intensity projections (MIP). The volunteers are ordered by weight, as in Tab. 5.1. Noteably, the local torso SAR shows a similar pattern in all volunteers. Higher SAR values are generally observed in the muscle segments rather than in fat segments. The maximum local SAR values occurred predominantly at joints (wrists, elbows, shoulders) and at the muscle attachment points at the pelvic floor and the iliac crest. All these locations represent regions where the muscle diameter decreases, leading to an increased local current density [18]. The corresponding numerical values are plotted in Fig. 5.3a and also listed in Tab. 5.2. The maximum local torso SAR tends to increase with body weight. By contrast, the maximum local arm SAR shows a relatively large variation and such a trend could not be found. Fig. 5.3a also shows the CV values of the B_1 field as a measure of the expected excitation performance.

Interestingly, all models in Fig. 5.2 show a noticeable asymmetry of the SAR distribution in left-right direction. To understand this effect, a transversal cross-section of volunteer 5 is shown in Fig. 5.4. The slice location was chosen at the dominant SAR hotspot at the iliac crest, representing an extreme case of this asymmetry. The top row of Fig. 5.4 shows the (a) water and (b) fat images used for model generation to give an overview of the local anatomy. The corresponding slice of the dielectric body model in Fig. 5.4c shows the voxel-by-voxel tissue classification, discriminating between low conductive, fatty tissues and highly conductive, water-rich tissues, where the latter represent the potential eddy current pathways in the body. In Fig. 5.4d, the maximum eigenvalues of the local Q-matrices are overlaid with the segmented model. These eigenvalues represent the largest possible SAR at each voxel for channel weights $\mathbf{w}$ with a unity norm. It can be seen that high SAR values can potentially occur in the arms and in the muscle tissues at the iliac crest (more specifically the cranial ends of the glutei medii muscles). Importantly, the eigenvalue distribution is approximately symmetric in the left-right direction, indicating that the channel weights $\mathbf{w}$ determine the level of asymmetry. At quadrature excitation inputs (cf. Fig. 5.4e and g), a diagonal pattern of the SAR distribution is observed in the torso, with elevated SAR values at one side of the belly and at the opposite side of the back. When performing RF shimming (cf. Fig. 5.4f and h), the SAR distribution becomes more homogeneous and the maximum SAR value is substantially reduced.

RF shimming without constraints

Fig. 5.2b and Fig. 5.2c show the SAR distribution for unconstrained RF shimming in all volunteers. The SAR pattern in the torso is often more

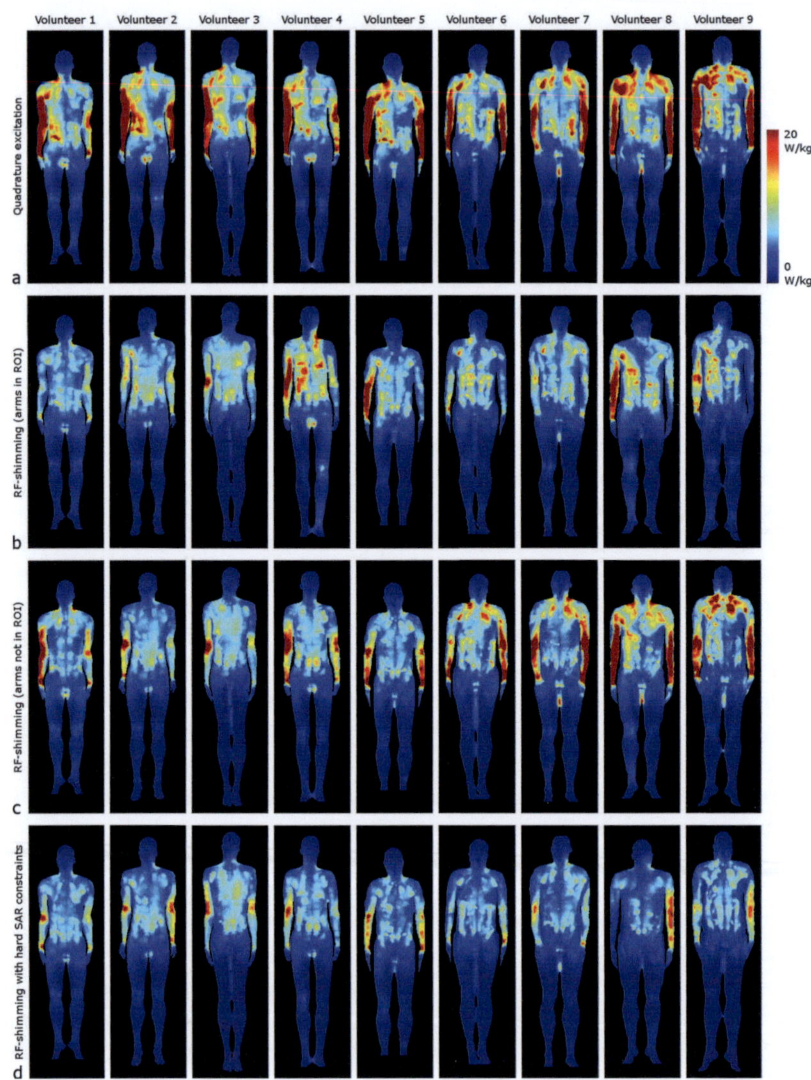

Fig. 5.2: MIPs of SAR (normalized to $2\,\mu T$): Compared to quadrature excitation (a), RF shimming (b-d) resulted in a significant SAR reduction. The SAR in the arms is lower if the arms are included in the ROI for shimming (b) than when excluding (c). With strict SAR constraints (d), the SAR is further reduced in the torso and in the arms.

RF excitation	Torso SAR [W/kg]			Extremity SAR [W/kg]			CV (B_1) [%]		
	min	max	mean	min	max	mean	min	max	mean
Quadrature	20.5	45.7	30.8	43.0	80.2	57.8	23.7	37.5	30.0
Unconstrained shim									
arms in ROI	13.2	22.7	17.4	14.8	34.0	23.2	11.2	23.6	14.7
arms NOT in ROI	12.0	28.3	18.7	27.5	59.7	37.2	11.0	16.6	12.2
SAR-constrained shim									
individual models	5.0	7.3	6.0	9.7	16.6	14.0	11.6	20.4	14.9
biggest model	6.2	8.3	6.9	8.7	20.1	14.2	13.3	18.2	15.8

Tab. 5.2: Summary of local SAR values and coefficients of variation (CV) of the B_1 field. All SAR values were calculated using the uncompressed Q-matrices.

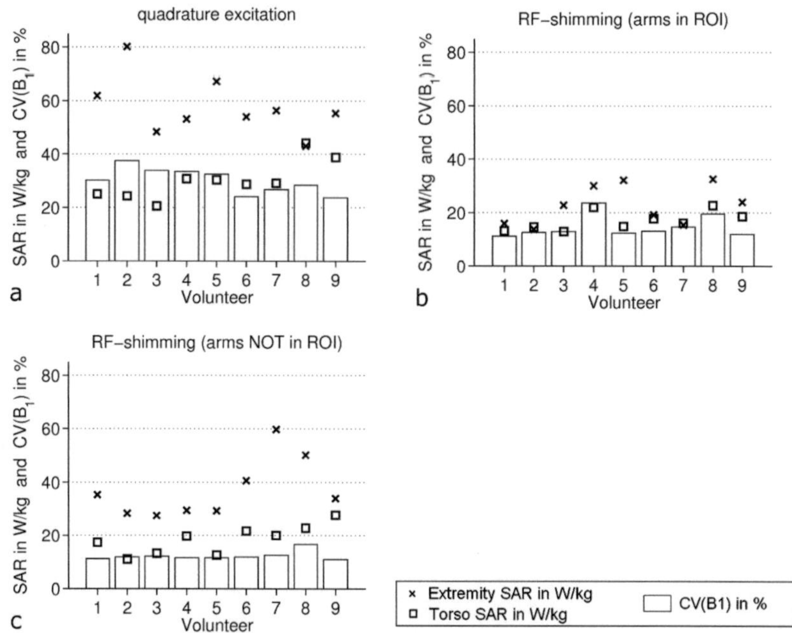

Fig. 5.3: Maximum local SAR and CV(B_1) values without SAR constraints: a) At quadrature excitation, the local torso SAR tends to increase with body weight. b) For unconstrained RF shimming (arms in the ROI), the maximum local torso and extremity SAR values are both substantially reduced and CV(B_1) is significantly improved. c) When excluding the arms from the shim ROI, the torso SAR values remain at a low level, whereas the SAR in the arms is elevated for some of the volunteers.

symmetric in the shimmed case and the maximum SAR in the arms is reduced (especially if these are inside the shimmed ROI).

This SAR-reducing effect of RF shimming was observed also quantitatively for all volunteers. First, the arms were included in the region-of-interest (ROI) used for the B_1 homogeneity optimization (c.f. Fig. 5.3b). In this case, the local SAR in the torso as well as in the arms was significantly reduced for all volunteers. The CV values demonstrate consistently improved RF field homogeneity.

Second, the arms were excluded from the ROI (c.f. Fig. 5.3c), optimizing B_1 homogeneity only in the torso. The achieved CV values were

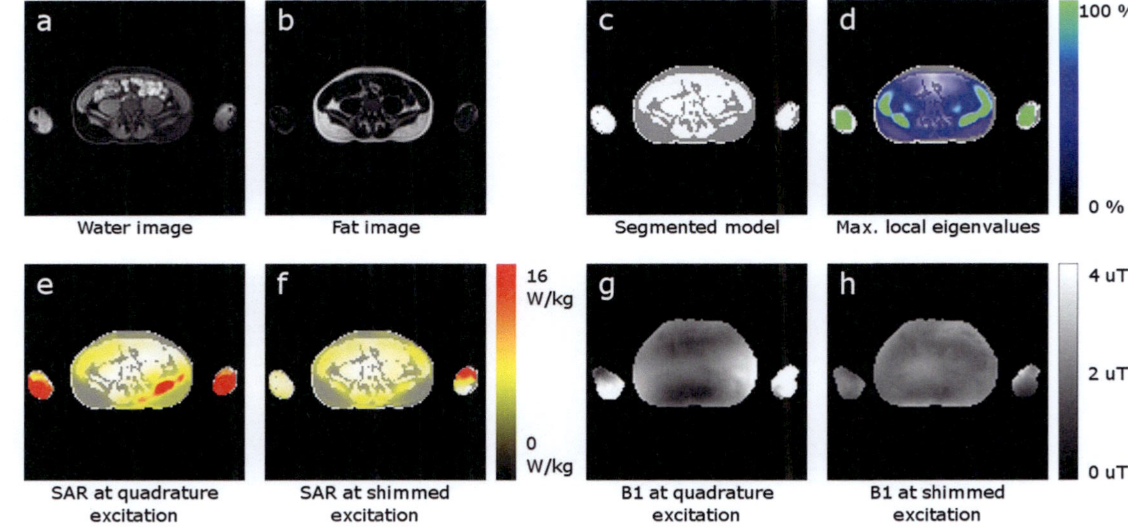

Fig. 5.4: Transverse cross-section through the model of volunteer 5: a) and b) Water and fat images. c) Segmented dielectric body model, white and gray represent water-rich tissues and fatty tissues, respectively. d) An overlay with the maximum eigenvalue of the local Q-matrices. This shows potential SAR hotspots occurring in the arms and in the torso in the muscles at the iliac crest at both sides of the body. e) At quadrature excitation, the actual SAR hotspots in the torso occur only at one side. f) With RF shimming, the SAR distribution is spatially more homogeneous and the maximum SAR is reduced. g) and h) The corresponding B_1 field (in the isocenter slice) illustrates the improvement of the excitation profile by RF shimming.

slightly better since the ROI was smaller. In that case, the local torso SAR was still reduced when compared to quadrature excitation. For the local SAR in the arms, this was no longer the case for all volunteers.

RF shimming with SAR constraints

The maximum local torso SAR for different SAR calculation approaches are plotted in Fig. 5.5a as a function of the "RF duty cycle" $B_{1,\mathrm{rms}}$ (cf. Eqs. 2.18 and 2.22) for one volunteer. In the worst-case SAR approach, the E field magnitudes were added. This overestimated the SAR at quadrature excitation by a factor of approximately seven. For unconstrained RF shimming, the SAR is reduced as reported in the previous section. For all these cases, SAR increases quadratically with $B_{1,\mathrm{rms}}$. With SAR constraints, the local torso SAR calculated with the compressed model remains strictly at the limit of $\mathrm{SAR}_{\mathrm{max}} = 10\,\mathrm{W/kg}$. The SAR calculated with the original model stays futher below the limit (at a margin roughly determined by $\varepsilon \cdot \mathbf{w}^H \mathbf{w}$). Fig. 5.5b shows that the maximum local extremity SAR is constrained similarly. Up to a demand $B_{1,\mathrm{rms}}$ of $1.2\,\mu\mathrm{T}$, the calculated SAR values rise quadratically and $\mathrm{CV}(B_1)$ is constant. Above this $B_{1,\mathrm{rms}}$ value, the SAR constraints take effect and the local SAR values remain below the respective limits. Instead, the B_1 field homogeneity is compromised. Up to about $2\,\mu\mathrm{T}$, this occurs only gradually. Above this value, the field homogeneity degrades rapidly but the SAR remains within the safety limits even for high values of $B_{1,\mathrm{rms}}$.

The SAR distribution for RF shimming with SAR constraints is shown in Fig. 5.2d and the relevant numerical values are given in Fig. 5.6a. An even further SAR reduction, compared to unconstrained RF shimming, can be observed and the local SAR maxima stay strictly below the chosen SAR limits, while the B_1 homogeneity (CV) remains in a similar range.

Fig. 5.6b shows the resulting local SAR values and $\mathrm{CV}(B_1)$ when using the SAR model of the biggest model (volunteer 9) instead of the individual SAR models. The B_1 field homogeneity is similar as in the previous case. The local SAR values are around the corresponding limits.

The time needed for the SAR-constrained RF shimming algorithm showed roughly a linear dependency on the number of local Q-matrices included. For a typical model with 250 local Q-matrices (after compression) the algorithm converged within 5 to 10 seconds on a desktop PC (C++ code on Windows XP, Intel Core 2 CPU at 2.4 GHz, 2 GB RAM).

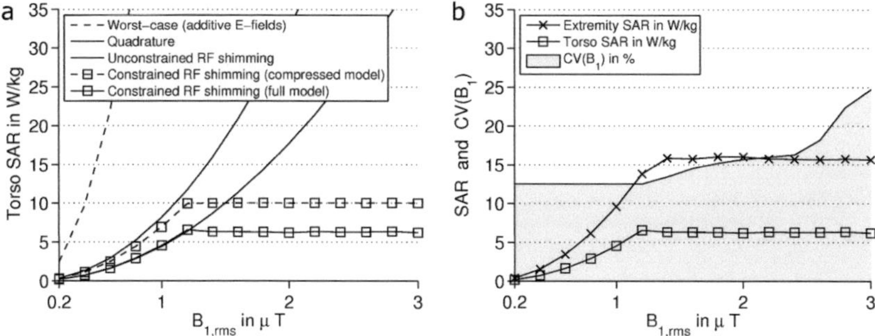

Fig. 5.5: SAR calculations with and without constraints (volunteer 7): a) The maximum local torso SAR is shown for different cases. Without constraints (worst-case, quadrature, unconstrained RF shimming), the maximum local torso SAR scales with the square of $B_{1,\mathrm{rms}}$. For constrained RF shimming, the SAR limit (10 W/kg) is reached at $B_{1,\mathrm{rms}} = 1.2\,\mu\mathrm{T}$. The SAR predicted by the compressed model closely approaches the limit. The actual SAR, calculated with the full model, stays strictly below the limit by a margin roughly proportional to ε. b) The SAR-constrained case is shown only. The local extremity SAR is restricted similarly as the local torso SAR. For low $B_{1,\mathrm{rms}}$, the coefficient of variation $\mathrm{CV}(B_1)$ is constant. As the SAR constraints become active, $\mathrm{CV}(B_1)$ increases, gradually at first but then faster. This clearly indicates the trade-off between RF shimming performance (uniformity) and the SAR. For comparison, $\mathrm{CV}(B_1)$ is 28.4% at quadrature excitation.

5.1.4 Discussion

At quadrature excitation, the SAR hotspots occurred at similar anatomical locations, typically towards the ends of long muscles, even though the volunteers represented a relatively broad range of adult male anatomies. The local torso SAR was found to increase for bigger patients. This trend was reported earlier in simulation studies when scaling the size of the Visible Human model [23]. This effect is due to the larger cross-sectional area for bigger volunteers that is penetrated by the B_1 field.

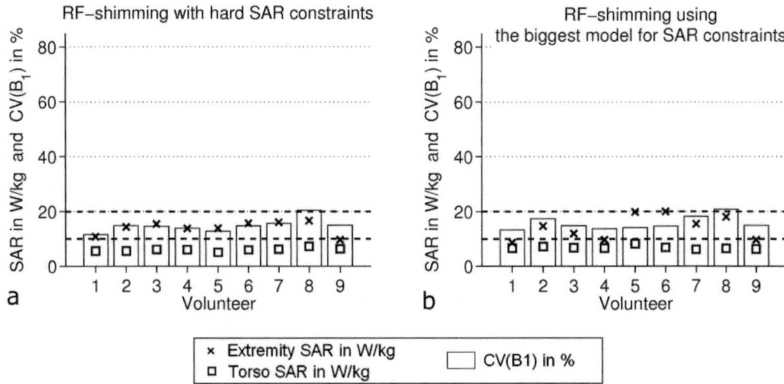

Fig. 5.6: Maximum local SAR and CV(B_1) values with SAR constraints: The dashed lines indicate the chosen local torso and extremity SAR limits (10 and 20 W/kg, respectively). a) If the personalized body model for each volunteer is used, the maximum local SAR values stay below the respective limits. b) If no personalized body model is available and the model of the biggest volunteer is used instead, the SAR values are in a comparable range but the limits are not necessarily strictly kept. All values are normalized to $B_1 = 2\,\mu\mathrm{T}$ in the isocenter slice.

This leads to an increased E field along the line integral around this area in Faraday's law and hence to an increase of the SAR. This increased dissipative loss also explains the higher RF power demand which is observed for bigger patients to achieve a certain B_1 level. For the SAR at the arms, such a trend could not be observed, which is possibly due to generally higher variations depending on slightly different positioning of the arms relative to the RF coil.

The SAR distribution in the torso showed a pronounced asymmetry which could be identified as a diagonal pattern in the transverse plane at quadrature excitation. Such a field distribution has also been described by van den Berg *et al.* [101] and has been attributed to the diagonal interference pattern of the electric field component resulting from the circularly polarized magnetic excitation inside an elliptically shaped body. Note that this elliptical pattern in the loaded coil deviates from the symmetric field pattern in an empty RF coil. In that study, it was demonstrated that RF shimming can effectively reduce this "coarse" electric field amplification in the human pelvis, even when the channel weights were optimized using an elliptical phantom for SAR constraints. Nevertheless, it was stated that local SAR hotspots are the result of very local anatomy-dependent field interactions [101]. Hence, the use of human body models is indispensable for accurate local SAR prediction.

In this study, RF shimming showed generally a reduction of the SAR for all volunteers, even when not taking any SAR constraints or regularization into account. Such a relation between improved B_1 homogeneity and reduced SAR was previously reported in a 2-channel body coil [102]. One explanation for this effect is that the dominant, clockwise-rotating B_1 field component becomes spatially more homogeneous. Since the induced E field follows from the spatial derivative, more precisely the curl, of the B_1 field according to Ampère's law, a smoother B_1 field will lead to a reduced E field. However, this SAR reduction may be limited to those regions where B_1 field homogeneity is demanded. For instance, when excluding the arms from the ROI, for two of the volunteers the local extremity SAR was slightly higher than at quadrature excitation, whereas the local torso SAR remained consistently reduced.

Care should however be taken when generalizing these findings for other types of RF coils and for other anatomies [112]. In practice, some regularization or penalty for the SAR will generally be advisable.

Using the complete SAR model, the SAR constrained algorithm requires several days, which is not practiceable. The model compression facilitated highly efficient SAR calculation and optimization. The proposed optimization method achieved this by a closed-form solution within a reasonably short time of a few seconds and yielded B_1 homogeneity values comparable to the unconstrained algorithm. The SAR values calculated with the uncompressed model were considerably below the chosen SAR limits. This indicates that the selected uncertainty margin of $\varepsilon = 20\%$ of the quadrature SAR is relatively large. Reducing the value to $\varepsilon = 10\%$ would however have increased the number of Q-matrices in the compressed model and hence the calculation time by a factor of roughly four. To compensate for this, the constrained RF shimming algorithm could be performed iteratively with decreasing values of ε.

Whereas this study focused on RF shimming, the barrier method can also be applied for the more general case of Transmit SENSE pulses. Due to the higher computational demand, it would then certainly be useful to reduce ε iteratively. The proposed approach can furthermore easily be extended by constraining the average and peak RF amplifier power on each channel as well as the global SAR.

Any simulation-based SAR management relies on the assumption that the actual patient is comparable to the model used for SAR estimation. One approach to cope with this situation is to use a "worst-case patient model" for SAR prediction. This was attempted in this section by performing RF shimming with the biggest model for SAR constraints. Choosing this model is reasonable as the torso SAR was found to increase with patient size and the arms of this model were closest to the transmit elements of the body coil. In this simulation, the SAR was found to be reduced for all volunteers compared to quadrature excitation and unconstrained RF shimming. This also indicates that the obtained channel weights probably lead to a favorable SAR distribution in the other models and will not lead to unexpected SAR hotspots. This is reasonable since all models used in this study were placed at the same position in the RF body coil and have a similar positioning of the arms.

In summary, RF shimming was found to substantially reduce the local SAR, consistently for all volunteers. To get a better handle on the local SAR, an efficient SAR-constrained algorithm was proposed and a further SAR reduction could be achieved with only minor compromises in RF performance.

5.2 A robust approach to SAR calculation

In the previous section, it was shown that different subjects can exhibit similar SAR hotspots which also tend to react similarly under multi-channel RF excitation for subjects in a comparable body position. In a clinical setting, a broad variety of patients has to be expected with respect to body position as well as age, gender and body size such that multiple sets of Q-matrices for various scenarios need to be prepared. In practice, it is not straightforward to choose the most appropriate body model. This section presents an alternative approach for efficient real-time calculation of the worst-case local SAR of a large number of different body models simultaneously for robust RF safety assessment.

Methods

The model compression approach according to Section 2.3.4 is applied twice to improve efficiency when using multiple body models. The algorithm consists of the following steps:

1. The model compression is performed for every single body model to reduce the computational effort of the following steps.

2. The local matrices $\mathbf{Q}$ of several models are combined to form a "robust model" which represents different anatomies and/or body positions. The maximum SAR calculated with this robust model represents an upper bound to the SAR of any included model.

3. The clustering algorithm is applied for a second time to further compress the robust model.

In this study, this multi-model clustering was performed for two cases: a) using the Visible Human Male at 18 positions of the patient bed along the bore axis of the MBC and b) using body models generated from 6 volunteers to represent different anatomies. The SAR overestimation term ε was defined as a percentage of the maximum local SAR at quadrature excitation.

Results and Discussion

The original number of SAR cells was 750,000 for the Visible Human Male and ranged from 310,000 to 540,000 for the volunteer models. The first clustering step (using $\varepsilon = 10\%$) reduced the number of cells in the Visible Human Male model to 83 – 5,190, depending on the position of the patient bed, and to 223 – 2,365 for the different volunteer models. The second clustering step resulted in a further reduction of the required number of cells, as shown in Fig. 5.7 as a function of ε. Using $\varepsilon = 10\%$ in the second clustering step yields a compression of the robust models by approximately 85% for both cases in addition to the first compression step without increasing uncertainty. This means that a considerable redundancy among the various models exists, as illustrated in Figs. 5.8 and 5.9. Further compression can be obtained by higher ε.

For the Visible Human Male, several positions of the patient bed were no longer included in the reduced generalized model. This means that the maximum local SAR does not occur at the respective position for any RF waveform. For the reduced generalized model generated from the volunteer models, such a relation was not found. This means that the maximum SAR can occur in any of the volunteers, depending on the applied RF waveform. Overall, the memory requirements as well as loading and SAR calculation times are strongly reduced.

The presented approach removes the redundancy between multiple body models. The calculation time is almost independent of the number of models used. This allows for efficient and robust local SAR estimation in parallel transmit MRI. The approach can be applied for real-time SAR calculations as well as SAR management.

5.3   Conclusions

RF shimming was found to substantially reduce the local SAR in 3 T parallel transmit body coil applications, even when not imposing any SAR constraints. The SAR burden can be further reduced by incorporating the SAR model directly into the RF shimming algorithm. Such a local SAR management can leverage previously SAR-limited MR sequences while simultaneously improving B_1 homogeneity without the need to trade off sequence timing or contrast parameters. In summary, RF shimming exhibits significant potential for SAR reduction.

In a clinical setting, patient-specific SAR models are usually not available. This chapter hence proposed a robust approach to represent different patient groups and body postures for comprehensive SAR prediction. The proposed approach is computationally efficient and requires only a

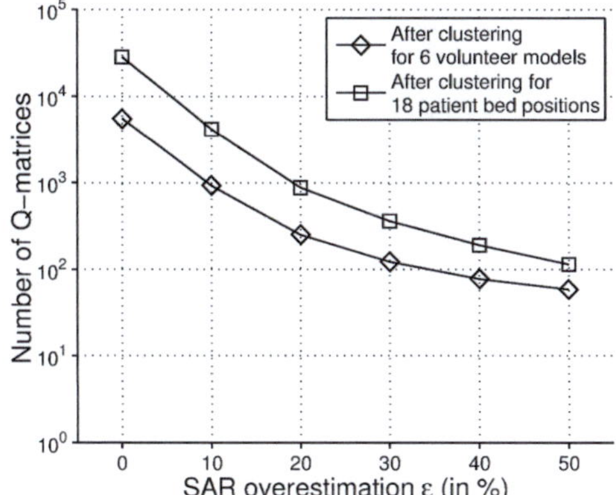

Fig. 5.7: Number of remaining Q-matrices as a function of the overestimation term ε in the second clustering step: When using the same value as in the first compression step ($\varepsilon = 10\%$), an additional compression of about 85% was observed for both cases investigated. This shows that there is considerable redundancy among the various models.

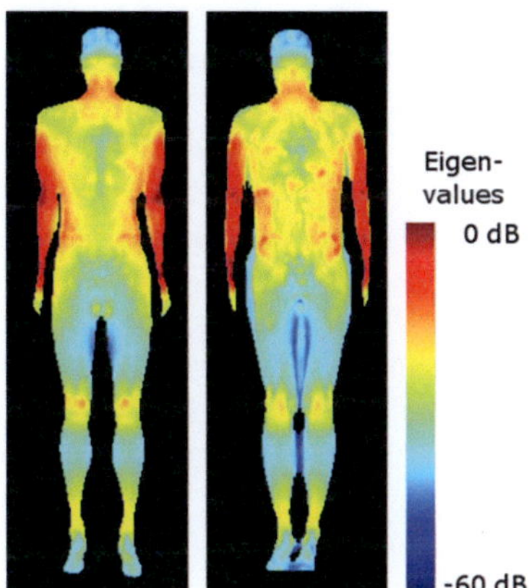

Fig. 5.8: Distribution of the maximum eigenvalues in volunteers 1 and 6 (maximum intensity projections): Despite the different anatomy, the potential for RF heating (i.e. the maximum eigenvalue distribution) is similar for both volunteers.

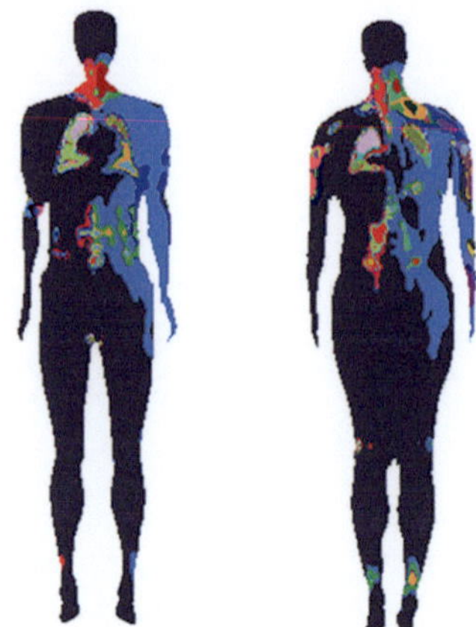

Fig. 5.9: Q-matrix clusters for the same two volunteers (coronal slice): Regions with the same color are represented by the same "dominating Q-matrix" after the second model compression step. The figure shows that these clusters are located at corresponding body regions in the two models. Whereas Fig. 5.8 only highlights that the potential SAR hotspots occur at similar locations, this figure confirms that both models respond similarly to a given vector of channel weights $\mathbf{w}$ as illustrated by the similarity of the local Q-matrices and the associated clusters. This explains the efficiency of the second clustering step to remove redundancy.

fractional additional effort compared to SAR calculation with a single model due to substantial redundancy among the different SAR models. Overall, this concept allows highly efficient SAR calculation and SAR management.

Chapter 6

SAR reduction by k-space adaptive RF shimming

Thus far, SAR calculations and SAR management have been discussed independently of the sampling process. In MR imaging, the RF Tx sequence is performed repeatedly for every profile in the sampling k-space. The center of the sampling k-space contains most of the signal energy [113] and determines image contrast and uniformity. This has been utilized for accelerated dynamic studies using the 'keyhole' concept [114, 115] and to reduce the average SAR of an MR scan by lowering the flip angle in the outer k-space for SAR critical protocols such as Steady State Free Precession (SSFP) [116] or hyperecho [117] sequences. Another example is to limit the relatively high-SAR magnetization transfer pre-pulses to the central k-space [118]. For parallel transmission, the k-space dependency offers an additional degree of freedom in the sequence design: Instead of using identical RF pulses for all sampling profiles, several different RF pulses can be applied [119].

In this chapter, a proof of principle for this approach is given for an RF shimming application, to achieve improved RF field homogeneity while simultaneously reducing SAR. The amplitudes and phases of each Tx-channel were adapted depending on the location in the sampling k-space. A phantom simulation study was performed for a better understanding of the additional freedom in the sequence design. Based on the simulation results, *in vivo* feasibility is shown for single-shot turbo spin echo (TSE) and fast field echo (FFE) imaging sequences.

Based upon: H. Homann, I. Graesslin, K. Nehrke, C. Findeklee, O. Dössel, and P. Börnert, "Specific Absorption Rate Reduction in Parallel Transmission by k-Space Adaptive Radiofrequency Pulse Design," *Magn. Reson. Med.*, vol. 65, pp. 350–357, 2011.

6.1 Methods

General concept

In parallel transmission, a desired excitation pattern can in general be achieved by various RF pulses. However, a trade-off exists between the RF field quality and the SAR as schematically shown in Fig. 6.1a. On this trade-off curve, a higher RF field performance is associated with higher SAR.

Considering the phase-encoding space of a Cartesian MR imaging sequence, different RF pulses could be chosen for each phase-encoding step of the image acquisition. However, for practical reasons it is straightforward to use the identical RF pulse for a group of phase-encoding steps with a similar distance to the k-space center, as illustrated in Figs. 6.1b-d. Image contrast and homogeneity are determined in the central k-space and hence, a maximally uniform RF field is desired here. An RF pulse that fulfills such high requirements may however cause relatively high SAR, if applied repeatedly for every k-space profile. Towards the outer sampling k-space, the homogeneity requirement can be relaxed which allows for application of RF pulses with lower power. This combination of different RF pulses facilitates a reduction of the average SAR of the entire scan.

Phase-consistent RF shimming

For the purpose of switching between different RF pulses during image acquisition, it is crucial that all k-space profiles are acquired with a similar Tx-phase in the object space to avoid phase-related artifacts such as ghosting. This section describes an approach to achieve such a phase-consistent excitation for RF shimming.

Recall, that the magnetic field $B_1^+(\mathbf{w}, \mathbf{r})$ is given by the superposition of the magnetic transmit field sensitivities $\mathbf{S}_{B1}(\mathbf{r})$ with the channel weights $\mathbf{w}$:

$$B_1^+(\mathbf{w}, \mathbf{r}) = \mathbf{S}_{B1}(\mathbf{r}) \cdot \mathbf{w} \tag{6.1}$$

For RF shimming, the shim setting $\mathbf{w} = (w_1, \ldots, w_N)$ is determined such that the RF field becomes optimally homogeneous. In many applications, the phase of the RF field is of little interest and only the magnitude is required to be homogeneous in the optimized RF field. For this purpose,

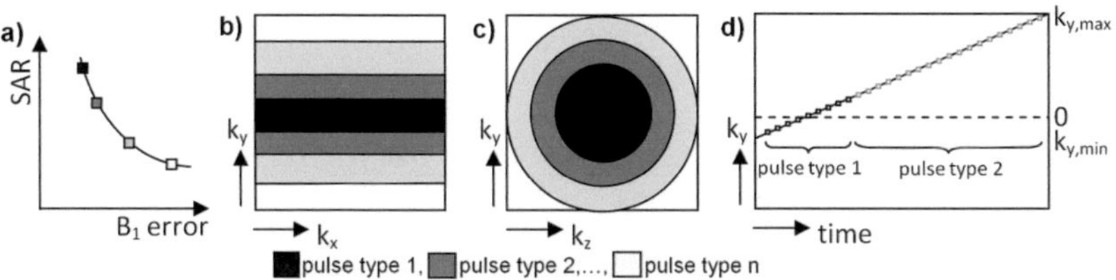

Fig. 6.1: General concept of k-space dependent RF pulse selection: a) Schematic of the trade-off between SAR and RF field error. Each of the squares on the curve represents a different RF pulse. Dark colored areas indicate an RF pulse with better performance, but requiring higher SAR. b) For a 2D sampling k-space (k_x: frequency encoding direction, k_y: phase encoding direction), the central phase-encoding profiles are acquired with RF pulses that have the least B_1 error but higher SAR. c) For 3D imaging, this can likewise be done in the two phase encoding directions k_y and k_z (the frequency encoding direction k_x is not shown here). d) A practical implementation is shown as a timing diagram for a single-shot 2D partial Fourier acquisition sequence with linear profile order. The first, central phase-encoding profiles in k_y are acquired with the high-quality pulse type and the outer profiles are acquired with a low power pulse type.

a magnitude least-squares (MLS) algorithm can be applied to solve the optimization problem without constraining the image phase [110, 120]:

$$\mathbf{w} = \arg \min_{\mathbf{w}} \{|| \, |B_1^+(\mathbf{w}, \mathbf{r})| - \hat{b}(\mathbf{r})||_2^2 + R(\mathbf{w})\} \qquad (6.2)$$

where $\hat{b}(\mathbf{r})$ denotes the real-valued target field and $R(\mathbf{w})$ is an appropriate regularization term. Compared to a simple least-squares (LS) optimization, this provides additional freedom in the optimization and permits improved magnitude profiles. The MLS optimization problem can efficiently be solved iteratively as linear equation systems by using the local variable exchange method [111].

The method proposed here represents a combination of the MLS and an LS approach: First, an MLS optimization step (Eq. 6.2) is performed to achieve a "high quality shim setting" $\mathbf{w}_{\mathrm{MLS}}$ for the central k-space. The resulting, but so far arbitrary, Tx-phase $\angle B_1^+(\mathbf{w}_{\mathrm{MLS}}, \mathbf{r})$ is then passed to an LS optimization step where it is used as the target phase:

$$\mathbf{w} = \arg \min_{\mathbf{w}}\{||B_1^+(\mathbf{w}, \mathbf{r}) - \hat{b}(\mathbf{r}) \cdot \exp(j\angle B_1^+(\mathbf{w}_{\mathrm{MLS}}, \mathbf{r}))||_2^2 + R(\mathbf{w})\} \quad (6.3)$$

This method will be referred to as "phase-consistent RF shimming". Using Eq. 6.3, a set of additional shim settings for different k-space areas can be calculated. The shim settings obtained in the LS step hence contain the optimized phase from the MLS step, but can be subject to a stricter regularization of the RF power and will hence be referred to as "low power shim-settings".

For simplicity, a conventional Tikhonov regularization is used for Eqs. 6.2 and 6.3 to control the RF input power $\mathbf{w}^H\mathbf{w}$:

$$R(\mathbf{w}) = \lambda \cdot \mathbf{w}^H\mathbf{w} \cdot \sum_{\mathbf{r}}\sum_{i}|S_i(\mathbf{r})|^2 \quad (6.4)$$

Here, λ is the penalty parameter and $\mathbf{w}^H$ denotes the Hermitian of the shim setting $\mathbf{w}$. The constant number $\sum_{\mathbf{r}}\sum_{i}|S_i(\mathbf{r})|^2$ is introduced to achieve λ values that are independent of the measured transmit field sensitivities $S_i(\mathbf{r})$; $\lambda = 1$ is typically a good starting value.

6.2 Phantom study

Experiments and simulations

All experiments were performed on the 3T Philips Achieva with the MBC for RF transmission. For signal reception, a 16-channel torso coil was used. The individual images were combined using a sum-of-magnitude reconstruction. A SENSE reconstruction was not applied in this study to avoid any potential influence of possible receive field inhomogeneities of the body coil.

The transmit sensitivities $S_{B1,i}(\mathbf{r})$ for each Tx-channel were acquired using the Actual Flip Angle technique (AFI). The protocol parameters were $TR_1 = 20\,\mathrm{ms}$, $TR_2 = 100\,\mathrm{ms}$, echo time $TE = 2.3\,\mathrm{ms}$, flip angle $60°$, field of view $480{\times}240{\times}60\,\mathrm{mm}^3$ at a scan resolution of $5{\times}5{\times}5\,\mathrm{mm}^3$.

To evaluate the performance of the approach for different parameter settings, phantom studies were carried out using a cylindrical water phantom (Ø40 cm, conductivity $\sigma = 0.05$ S/m, permittivity $\varepsilon_r = 63$).

B_1 maps were acquired for each Tx-element and a high quality shim setting was calculated according to Eq. 6.2 using a weak regularization of the RF power to achieve a high field homogeneity. Then, seven phase-consistent shim settings were calculated according to Eq. 6.3 with increasing values of the power penalty parameter λ.

To simulate the RF field homogeneity of the k-space dependent combination of shim settings, the B_1 maps were transformed into k-space. A hybrid image was generated by combining profiles from the high quality data in the k-space center with profiles from the low power data in the outer k-space, ordered by increasing λ (cf. Fig. 6.1b). The hybrid data was then transformed back into the image space. This kind of simulation mimics an imaging sequence using k-space dependent selection of shim settings with uniform receive sensitivities at low flip-angles, where the image intensity is directly proportional to the RF field. Two types of simulations were carried out: First, the high quality shim setting was combined with the seven low power shim settings at random k-space fractions (ordered by increasing λ). This is referred to as the "Monte-Carlo simulation". Second, only two different shim settings were combined (the high quality shim setting and the shim setting with the lowest power) and the k-space fraction of the low power shim setting was increased from 0% to 100% in steps of 10%.

Results

Selected B_1 maps from the phantom experiment are shown in Fig. 6.2. The RF field in quadrature mode in Fig. 6.2a shows excessive spatial variation due to interference, since the RF wavelength (30 cm) is comparable to the phantom diameter. As expected, the B_1 map resulting from the high quality shim (see Fig. 6.2b) is relatively homogeneous, whereas the low power shim setting shows a strong interference pattern (see Fig. 6.2c). As required, the phase in both images shows only relatively small deviations of less than 15° at the phantom edges. The two B_1 maps shown in Figs. 6.2b and c were combined in k-space using 20% of the high quality profiles and 80% of the low power profiles, with the resulting map shown in Fig. 6.2d. The homogeneity is comparable to the pure high quality shim and no artifacts due to phase inconsistencies are visible.

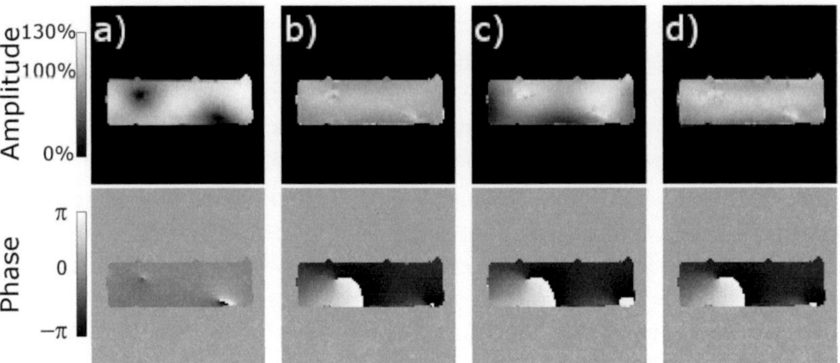

Fig. 6.2: B_1 maps based on phantom measurements (top: amplitude as percentage of the nominal flip angle, bottom: phase): a) RF field in quadrature mode, b) high quality MLS shim with little power regularization, c) phase-consistent shim setting with the maximum power regularization, and d) simulated k-space combined B_1 map (20% of the high quality shim, 80% of the phase-consistent low power SAR shim).

For quantitative evaluation of the phase-consistent RF shimming method, the coefficient of variation (CV) of the resulting B_1 maps and the corresponding average RF power are shown in Fig. 6.3a. As a reference, the quadrature mode is also shown. For simple LS shimming (using a target phase of $0°$ for every pixel) a trade-off curve was obtained by varying the regularization parameter λ. A significant reduction of RF power as well as of the B_1 error relative to the quadrature mode was achieved. MLS shimming performs even better since an arbitrary image phase is allowed. The proposed phase-consistent RF shimming approach requires slightly more RF power for a given CV than the MLS optimization due to the additional constraint to the excitation phase. As expected, both significantly outperform the conventional LS shim. The general trade-off is obvious for all cases, supporting the heuristic relationship shown in Fig. 6.1a.

Results for the k-space dependent shim switching mode are shown in Fig. 6.3b. The Monte-Carlo simulation shows that combining several shim-settings in k-space yields better results than using just one shim setting, i.e. than MLS shimming or phase-consistent RF-shimming without k-space switching. Interestingly, the combination of only two shim settings (the highest quality MLS-optimized shim setting and the shim setting with the lowest power) delineates nearly the optimal combination, as highlighted by the black curve. This combination achieves the RF field homogeneity that is provided by the high quality shim setting used in the central k-space at a maximum RF power reduction.

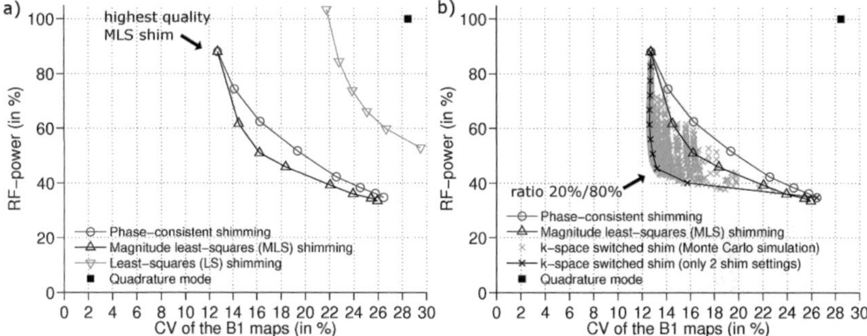

Fig. 6.3: Trade-off curves for RF shimming from the phantom study: a) Phase-consistent RF shimming performs comparable to the phase-unconstrained MLS design. By contrast, the LS design with a fixed target phase (0°) results in a less favorable solution, much closer to the quadrature mode. b) The k-space dependent combination of different phase-consistent shim settings in a Monte-Carlo manner pushes the achievable optimum further. In this example almost the optimal results can be achieved by combining just two different shim-settings (the highest quality MLS-optimized shim setting and the phase-consistent shim setting with the lowest power).

6.3 *In vivo* study

Experiments

To investigate *in vivo* feasibility, experiments were performed on seven healthy volunteers (all male, age $29-54$ years). Imaging was conducted in the abdominal region using two different imaging protocols: 1) a low flip-angle 3D FFE sequence where the effect of the RF field inhomogeneities is directly visible as the intensity is proportional to B_1 and 2) a more clinically relevant single-shot 2D TSE sequence where an inhomogeneous RF field can result in local signal loss due to the high number of refocusing pulses. Both protocols had a field of view of $450\times300\,\mathrm{mm}^2$ and a scan resolution of $1.4\times1.4\times5\,\mathrm{mm}^3$. The number of slices was 15 in the 3D FFE protocol and 3 in the 2D TSE protocol. For both protocols, a "shim switching mode" was implemented in the scanner software to switch between two different shim settings $\mathbf{w}$, depending on the k-space position during the scan. The scanner's standard RF waveforms (Sinc-Gaussian pulse shapes) were used for all experiments.

In practice, the profile order needs special consideration. For steady-state imaging, it takes a time interval roughly in the order of T1 after switching shim settings to allow the new steady-state to settle. In this study, 3D Cartesian sampling with an elliptic centric profile order was chosen for the FFE protocol. Sampling was started in the center of the 2D phase-encoding space (cf. Fig. 6.1c) and continued in an outwards spiral. After the central profiles had been sampled, the shim setting was switched and the trajectory was continued at the outmost profile and spiraled inwards to the location where the first spiral had ended. In this manner, a single switching event is required, minimizing any disturbance of the steady-state without a need for dummy RF pulses in the transition period. Image contrast was proton-density weighted using a flip angle of $4°$, TR $= 4.6\,\mathrm{ms}$, and an in-phase echo time TE $= 2.3\,\mathrm{ms}$.

For single-shot TSE sequences, sampling would typically need to be started in the k-space center with the high quality shim setting before the spin phase pathways are impaired by the less homogeneous RF field of the low power shim setting. In this study, a linear profile order with partial Fourier encoding was applied as illustrated Fig. 6.1d with a halfscan factor of 55%. Image contrast was T2-weighted by choosing TE $= 80\,\mathrm{ms}$ at an echo spacing of $4.9\,\mathrm{ms}$ and a TSE factor of 108. The nominal excitation and refocusing angles were $90°$ and $120°$, respectively.

Images were acquired during breathhold using four different settings: 1) in quadrature mode, 2) with an MLS-optimized high quality shim setting, 3) with a phase-consistent low power shim setting, and 4) in the "shim switching mode", combining the former two shim settings. As parameters, the regularization penalties and the switching point in k-space offer various combinations. For the high quality shim setting, the regularization was chosen such that the RF power was approximately equal to the power in the quadrature mode using $\lambda = 0.05$ for all volunteers. For the low power shim setting, the regularization was chosen such that the RF power was strongly reduced ($\lambda = 5.0$). After the optimization, all shim settings were rescaled such that the nominal flip-angle was achieved on average over the resulting B_1 map. A fraction of 20% of the profiles was empirically chosen for the high quality shim setting in the central k-space and the low power shim setting was used for the remaining 80%. In a blind evaluation, image quality was assessed independently by two reviewers on a 5-point scale (1: poor image quality; 5: good image quality). SAR calculations were carried out as described in Chapter 2, using the fully segmented Visible Human Male as a generic patient model.

Results

Images of the volunteer study for the different shimming strategies applied to both protocols are compared in Figs. 6.4 and 6.5. The quadrature mode images and the low power mode images show considerable local signal loss due to RF field inhomogeneity. These are patient-dependent and often localized around the spine which agrees with the pattern of the RF field. This is shown exemplarily for volunteer 6 for comparison of the images (Figs. 6.4 and 6.5, last row) with the corresponding B_1 maps (Fig. 6.6). The high quality shim settings improved image uniformity in all cases, which can also be observed in the shim-switched images.

Means and standard deviations from the blind evaluation for the different shim settings were as follows: quadrature mode (FFE images: 2.1 ± 1.1, TSE images: 2.2 ± 1.2), high quality shimmed (FFE: 3.7 ± 0.7, TSE: 3.9 ± 0.8), low power shimmed (FFE: 2.5 ± 1.1, TSE: 2.4 ± 1.0), and k-space-switched shimming images (FFE: 3.5 ± 1.0, TSE: 3.6 ± 0.7). The corresponding paired t-test showed that the high quality shimmed and the k-space-switched shimming images were rated significantly better than the quadrature mode and low power shimmed images ($p \leq 0.01$ in all cases). The differences between the high quality shimmed and the k-space-switched shimming images were found to be not significant ($p = 0.55$ for the FFE images and $p = 0.27$ for the TSE images).

The resulting RF power values are compared to the corresponding SAR estimates from the FDTD simulations in Tab. 6.1. Using the proposed k-space adaptive technique, RF power could be reduced to 59%–77%, relative to quadrature mode. The whole-body SAR and maximum local 10 g SAR were reduced to 50%–75% and 42%–73%, respectively. Furthermore, a high correlation was found between the RF power and the body-averaged SAR (Pearson's $r = 0.91$) as well as the local 10 g SAR ($r = 0.87$).

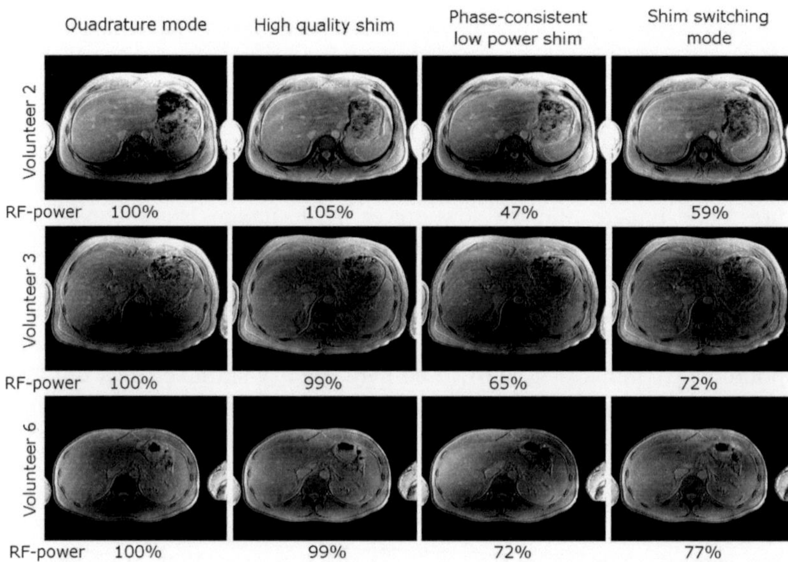

Fig. 6.4: *In vivo* 3D low tip angle FFE images: The intensity is directly proportional to the excitation field strength. The quadrature mode as well as the low power shim setting show areas of reduced and increased flip angle, varying from patient to patient. As expected, these field inhomogeneities can be resolved by RF shimming, as shown by the high quality shim images. The image homogeneity obtained by k-space dependent switching of shim settings is comparable to the high quality shim, however at a significantly reduced RF power (values given as percentages of the quadrature mode).

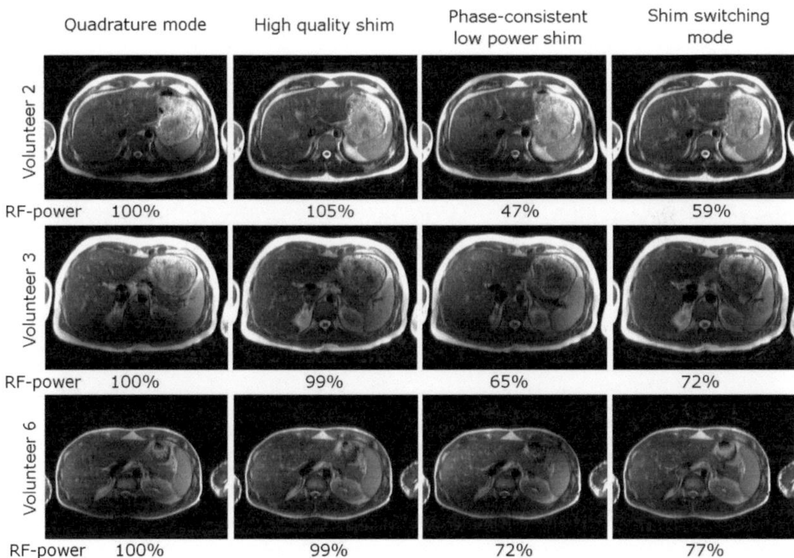

Fig. 6.5: *In vivo* 2D TSE images: As opposed to the FFE images in Fig. 6.4, the effect of RF field inhomogeneities on the image contrast is visible here. Both, the quadrature mode images and the low-power mode images, exhibit locally degraded contrast in areas of reduced flip angle. The images obtained with high quality shim settings and in the shim switching mode do not suffer from this effect. The relative RF power values are identical to the FFE sequence.

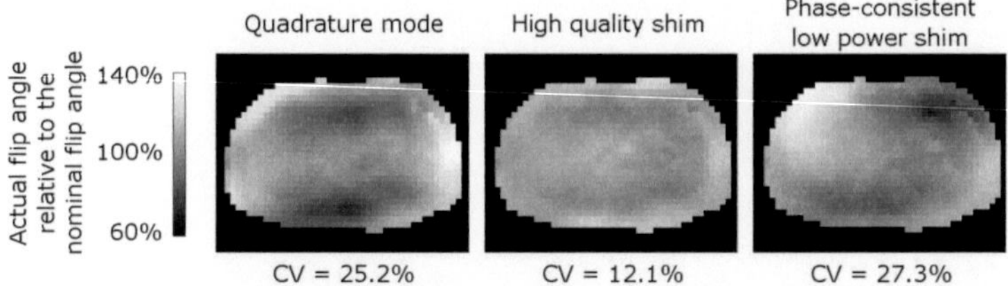

Fig. 6.6: B_1 maps based on *in vivo* measurements (volunteer 6). The quadrature mode (left) shows a highly inhomogeneous pattern, often exhibiting a low amplitude near the spine. The MLS-optimized map (center) does not show this effect and has the lowest coefficient of variation (CV). The LS-optimized map (right) has the lowest RF power (72% of the quadrature mode) and exhibits a CV in the range of the quadrature map, however at a differently distributed interference pattern.

6.4 Discussion

Feasibility of k-space dependent RF shimming was demonstrated *in vivo* and on phantom data. In comparison to quadrature mode, image contrast and homogeneity were significantly improved as in conventional RF shimming, but at a significantly lower SAR. The phase-consistent RF shimming approach allowed for combining different shim settings without introducing phase-related artifacts in the images.

The simulation study confirmed that the phase-consistent optimization performed notably better than a simple least-squares optimization, allowing for a higher SAR reduction. Furthermore, it can be concluded from the phantom study that a limited number of shim settings is sufficient to find an appropriate compromise between RF shimming performance and SAR. Choosing only two different shim settings appears to be a straight-forward and effective combination that might give reasonably good results for many applications. Using more than two shim settings, however, may be advantageous in some cases, e.g. to achieve a smooth transition in k-space when switching in steady-state imaging.

Consequently, the *in vivo* study was conducted with only two different shim settings. Switching between different shim settings imposes some

Volunteer	High quality shimming			Low SAR shimming			Switched shimming		
	RF power	body SAR	10 g SAR	RF power	body SAR	10 g SAR	RF power	body SAR	10 g SAR
1	112%	136%	116%	64%	58%	50%	74%	74%	55%
2	105%	97%	90%	47%	38%	30%	59%	50%	42%
3	99%	111%	107%	65%	61%	56%	72%	71%	64%
4	110%	115%	99%	61%	65%	64%	71%	75%	71%
5	96%	93%	97%	63%	60%	44%	69%	66%	49%
6	99%	94%	81%	72%	64%	48%	77%	70%	49%
7	100%	110%	130%	58%	65%	59%	66%	74%	73%
Average	103%	108%	103%	61%	59%	50%	70%	69%	58%

Tab. 6.1: RF power and SAR values for all volunteers. The values apply to the FFE sequence as well as to the TSE sequence as the identical shim settings were used for both protocols. All values are given as percentages relative to the quadrature mode so that the RF waveform and the sequence timing does not affect the numbers.

constraints on the design of the profile order, as the number of switching events needs to be minimized to reduce the influence on the steady-state or on the spin phase pathways. In this study, suitable profile orders for 3D FFE and 2D TSE imaging were developed, such that image contrast was maintained. The low flip-angle FFE images showed a direct relation to the B_1 field distribution and demonstrated that homogeneity could be significantly improved by k-space adaptive shimming. The T2-weighted TSE images showed that image contrast was similarly improved.

As opposed to the previous section, the SAR was not necessarily reduced when using only a weak regularization for "high quality shimming". This is probably because the Visible Human was used as a generic SAR model in this study. However, a stronger regularization of the RF input power resulted in a significant reduction of body-averaged and maximum local 10 g SAR.

The parameters used for the *in vivo* study were chosen empirically here. Obviously, there is further potential in optimizing these parameters with respect to acceptable local errors of the flip angle and the transmit phase. It is important to note that such optimization can be performed on-line based on the B_1 maps prior to actual imaging. To maintain SAR limits, the approach should be combined with a SAR management scheme as proposed in the previous chapter.

The k-space dependent RF pulse selection can in principle be combined with all kinds of existing SAR reduction and RF pulse optimization techniques. For example in TSE sequences, a joint design [121] of the excitation and several refocusing pulses could prove useful to locally control accumulating effects of imperfections in the refocusing pulses. In hyperecho sequences [117], the 180° core RF refocusing pulse requires outstanding homogeneity and hence a high quality shim setting with a higher SAR could be chosen for this specific pulse whereas the remaining refocusing pulses are less crucial and could be used for k-space adaptive SAR reduction.

Furthermore, the k-space dependent parallel transmit approach may likewise be applied to all kinds of magnetization preparation pulses. For example, outer volume suppression pulses can significantly contribute to the SAR of an MR sequence and are hence well suited to SAR reduction by parallel transmission. Many preparation pulses do not require phase consistency and a normal magnitude least-squares optimization could be used. Overall, there is a wide field of applications which can benefit from the k-space dependent parallel transmit sequence design.

6.5 Conclusions

A new approach to SAR reduction in parallel transmission MR systems based on k-space dependent RF pulse selection was presented. The method allows to achieve an image homogeneity and contrast comparable to magnitude least-squares shimming with additionally reduced average SAR. The approach supports improved image contrasts in high-field MR imaging and spectroscopic imaging and potentially allows for reduced examination times in otherwise SAR-limited protocols. In principle, the proposed concept is applicable to RF shimming as well as to the general Transmit SENSE framework and can be combined with SAR management concepts.

Chapter 7

Summary

The present thesis contributes new approaches for comprehensive SAR prediction to ensure safe operation of parallel transmit MRI. This furthermore bridges the gap to control the SAR to maintain safety limits while simultaneously optimizing excitation performance, which is referred to as *SAR management*.

The thesis first focused on modeling of a multi-channel RF transmit coil and its cross-calibration to the physical multi-channel RF system. The simulations were validated by measurements of the electric and magnetic fields inside the MR scanner as well as by temperature measurements to estimate the total dissipated power.

Reliable modeling of the patient body is generally challenging, as patients not only vary in body size and composition as well as body fat content but are also examined at different body positions within the MR scanner. To account for these variations, multiple body models are required to cover the complete range of different patients and body positions. In this work, a novel approach to generate reasonably accurate body models based on water-fat separation is proposed. Based on this, a first *in vivo* validation of RF field simulations by means of B_1 mapping was achieved. In a clinical setting, it is currently not practical to generate individualized body models for every patient. Instead, the model generation approach can be used to generate a library of body models, representing different anatomies and body positions.

Moreover, this work has proposed an efficient and robust approach for prediction of the maximum local SAR that can occur in any of multiple models for a given set of transmit waveforms, based on a clustering approach to remove the redundancy between the models. This allows for a worst-case SAR prediction when the actual patient anatomy and position is not known and can be applied to generate a comprehensive model for routine use.

For the future, the bottleneck of long simulation times might be overcome by advanced EM simulation techniques, allowing even more accurate safety assessment by true patient-specific models. Another interesting option is that the magnetic field sensitivities might also be obtained from such simulations to avoid the need for B_1 mapping.

Beyond SAR prediction, this work investigated the potential of SAR management by parallel transmission with a focus on RF shimming. It was shown in a broad variety of human body models that RF shimming tends to reduce SAR. To ensure that the SAR remains within regulatory limits while simultaneously increasing Tx field homogeneity, an efficient approach for RF shimming with strict local SAR constraints was developed, based on the barrier method. This approach can employ the model used for SAR prediction at the MR scanner and thus does not require additional assumptions.

Finally, a novel approach for SAR reduction by k-space adaptive RF shimming was developed. Relatively SAR-intense, but high-quality RF pulses are applied in the k-space center, whereas low-SAR pulses are applied in the outer k-space. Feasibility of this concept was demonstrated for FFE and TSE sequences. A notable SAR reduction with optimized image contrast was achieved with marginal compromises on image quality. This approach can be combined with SAR management techniques to further minimize SAR.

Further work is required before the mentioned approaches can be broadly employed. This should include the generation of a more complete model library, e.g. representing children, females, obese people, and patients with pathologies in various body positions. Furthermore, the choice of models representing the actual patient and the suitable descriptive parameters requires further investigation. Moreover, SAR management in parallel transmit applications beyond RF shimming has large further potential. Also the application of this work to MRI at ultra-high field strengths ($\geq 7\,\mathrm{T}$) needs further discussion.

To conclude, the SAR is a major limiting factor for high-field MRI and requires extensive consideration for parallel transmission. Parallel transmission also offers a new degree of freedom to control the SAR. By accurate modeling of the multi-channel RF coil and of the patient body therein, a substantial SAR reduction can be achieved. This facilitates MR sequences which would otherwise not be applicable due to SAR limits in addition to improved excitation performance. Overall, parallel transmission with a suitable SAR management allows improving the diagnostic value of MRI by better contrasts as well as scan efficiency, for the benefit of clinicians and patients.

Appendices

A.1 Maxwell's equations

As Maxwell's equations are cited repeatedly throughout this work, the equations are briefly listed here in terms of the electric field E and the magnetic flux density B for linear and isotropic materials. The electric displacement field D, the magnetic field strength H, and the free current density J can be found by substituting the material relations $D = \varepsilon_r \varepsilon_0 E$, $J = \sigma E$, and $B = \mu_0 H$.

	Differential form	**Integral form**
Ampère's law	$\operatorname{rot}\frac{1}{\mu_0}B = \varepsilon_0\varepsilon_r\dot{E}+\sigma E$	$\oint\frac{1}{\mu_0}B\,\mathrm{d}s = \int(\varepsilon_0\varepsilon_r\dot{E}+\sigma E)\mathrm{d}A$
Faraday's law	$\operatorname{rot}E = -\dot{B}$	$\oint E\,\mathrm{d}s = -\int\dot{B}\,\mathrm{d}A$
Gauss's law for the electric field	$\operatorname{div}\varepsilon_r\varepsilon_0 E = \rho_{el}$	$\oiint\varepsilon_r\varepsilon_0 E\,\mathrm{d}A = \int\rho_{el}\,\mathrm{d}V$
Gauss's law for the magnetic field	$\operatorname{div}B = 0$	$\oiint B\,\mathrm{d}A = 0$

A.2 Decomposition of the B_1 field to polarized components

A magnetic field with a right-handed polarized component B_1^+ and a left-handed polarized component B_1^- is shown in Fig. A.1. Using two small pick-up coils, oriented in x- and y-direction, two time-signals can be detected in the laboratory frame:

$$
\begin{aligned}
b_{1,x}(t) &= \hat{B}_1^+ \cos(\omega t + \phi_+) + \hat{B}_1^- \cos(\omega t + \phi_-) & \text{(A.1)}\\
b_{1,y}(t) &= \hat{B}_1^+ \sin(\omega t + \phi_+) - \hat{B}_1^- \sin(\omega t + \phi_-) & \text{(A.2)}\\
&= \hat{B}_1^+ \cos(\omega t + \phi_+ - \frac{\pi}{2}) - \hat{B}_1^- \cos(\omega t + \phi_- - \frac{\pi}{2}) & \text{(A.3)}
\end{aligned}
$$

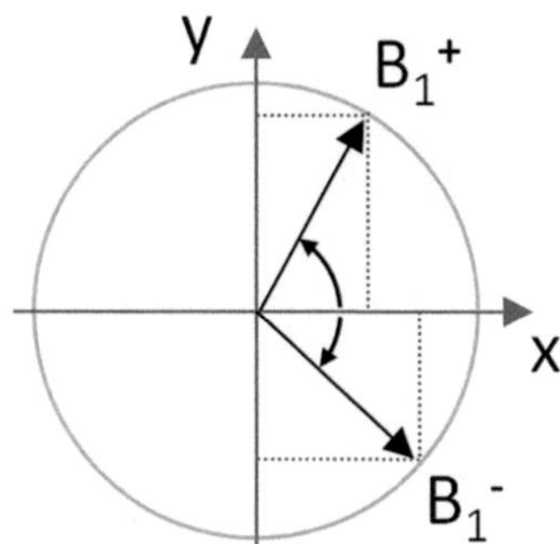

Fig. A.1: Relation of the rotating B_1 field components in the laboratory frame: The magnetic fields B_1^+ and B_1^- rotate at ωt in counter-clockwise and clockwise direction, respectively. The observed time-signals $b_{1,x}(t)$ and $b_{1,y}(t)$ follow from the projections onto the x- and y-axis.

Here, the circumflex denotes the amplitude of the B_1 phasors and $\phi_\pm$ denotes the associated phase. These time-signals can be considered as the real parts of a complex representation in the frequency domain:

$$
\begin{aligned}
b_{1,x}(t) &= \Re(B_{1,x} e^{i\omega t}) & \text{(A.4)}\\
b_{1,y}(t) &= \Re(B_{1,y} e^{i\omega t}) & \text{(A.5)}
\end{aligned}
$$

where the phasors in the laboratory frame $B_{1,x}$ and $B_{1,y}$ are defined by the complex expansion of the time-signals:

$$
\begin{aligned}
B_{1,x}e^{j\omega t} &= \hat{B}_1^+\left(\cos(\omega t + \phi_+) + i\sin(\omega t + \phi_+)\right) \\
&\quad + \hat{B}_1^-\left(\cos(\omega t + \phi_-) + i\sin(\omega t + \phi_-)\right) \\
&= B_1^+ e^{i\omega t} + B_1^- e^{i\omega t}
\end{aligned}
\tag{A.6}
$$

$$
\begin{aligned}
B_{1,y}e^{j\omega t} &= \hat{B}_1^+\left(\cos(\omega t + \phi_+ - \frac{\pi}{2}) + i\sin(\omega t + \phi_+ - \frac{\pi}{2})\right) \\
&\quad - \hat{B}_1^-\left(\cos(\omega t + \phi_- - \frac{\pi}{2}) + i\sin(\omega t + \phi_- - \frac{\pi}{2})\right) \\
&= B_1^+ e^{i(\omega t - \pi/2)} - B_1^- e^{i(\omega t - \pi/2)} \\
&= -iB_1^+ e^{i\omega t} + iB_1^- e^{i\omega t}
\end{aligned}
\tag{A.7}
$$

which can be written in matrix form:

$$
\begin{pmatrix} B_{1,x} \\ B_{1,y} \end{pmatrix} = \begin{pmatrix} 1 & 1 \\ -i & i \end{pmatrix} \begin{pmatrix} B_1^+ \\ B_1^- \end{pmatrix}
\tag{A.8}
$$

Inverting the matrix yields an expression for calculating the rotating field components from the description in the laboratory (or simulation) frame:

$$
\begin{pmatrix} B_1^+ \\ B_1^- \end{pmatrix} = \frac{1}{2} \begin{pmatrix} 1 & i \\ 1 & -i \end{pmatrix} \begin{pmatrix} B_{1,x} \\ B_{1,y} \end{pmatrix}
\tag{A.9}
$$

Similar relations are used in the Jones vector notation, commonly applied in optics to described circularly polarized light. However, the Jones formalism typically uses the electric fields and a different normalization. Note that a phasor description of the B_1 components is used (i.e. a right-hand rotating reference frame for B_1^+ and for B_1^-), whereas Hoult [122] defines B_1^- in a left-hand rotating reference frame such that his definition of B_1^- is the complex conjugate of the definition used in this work. However, both descriptions are consistent with each other as Hoult uses a second conjugation in his equation for the received MR signal, canceling the first. The advantage of the definition used here is that the equations are completely linear.

A.3 Expectation-maximization algorithm

To model the water- and fat-dominated tissues, two Gaussian distributions were used (Eqs. A.10 and A.11). An exponential distribution (Eq. A.12) was chosen to model the background noise. The mixture model was defined as the sum in Eq. A.13, with weighting factors $a_w + a_f + a_b = 1$. In these equations, x denotes the image intensity and $\Theta = \{a_w, \mu_w, \sigma_w; a_f, \mu_f, \sigma_f; a_b, \mu_b\}$ is the set of model parameters.

$$p_w(x|\mu_w, \sigma_w) = \frac{1}{\sqrt{2\pi}\sigma_w} \exp\left(-\frac{(x-\mu_w)^2}{2\sigma_w^2}\right) \tag{A.10}$$

$$p_f(x|\mu_f, \sigma_f) = \frac{1}{\sqrt{2\pi}\sigma_f} \exp\left(-\frac{(x-\mu_f)^2}{2\sigma_f^2}\right) \tag{A.11}$$

$$p_b(x|\mu_b) = \frac{1}{\mu_b} \exp\left(-\frac{x}{\mu_b}\right) \tag{A.12}$$

$$p(x|\Theta) = a_w \cdot p_w(x|\mu_w, \sigma_w) + a_f \cdot p_f(x|\mu_f, \sigma_f) + a_b \cdot p_b(x|\mu_b) \tag{A.13}$$

The optimal model parameters can be found iteratively by an Expectation-Maximization algorithm, similar to [95]. In the expectation step, the model parameters are given and the probability that a voxel of intensity x belongs to either segment is calculated as:

$$E_w(x) = \frac{a_w \cdot p_w(x|\mu_w, \sigma_w)}{p(x|\Theta)} \tag{A.14}$$

$$E_f(x) = \frac{a_f \cdot p_f(x|\mu_f, \sigma_f)}{p(x|\Theta)} \tag{A.15}$$

$$E_b(x) = \frac{a_b \cdot p_b(x|\mu_b)}{p(x|\Theta)} \tag{A.16}$$

In the maximization step, the model parameters Θ are updated based on the observed histogram $H(x)$ by maximizing the log-likelihood function $\mathcal{L}$:

$$\Theta' = \arg\max_{\Theta} \mathcal{L}(\Theta|H(x)) \tag{A.17}$$

Solving Eq. A.17 for the optimal parameters gives the following expressions, where the histogram $H(x)$ is weighted with the terms obtained in the expectation step:

$$a'_w = \frac{1}{N} \sum_i H(x_i) \cdot E_w(x_i) \tag{A.18}$$

$$\mu'_w = \frac{\sum_i x_i \cdot H(x_i) \cdot E_w(x_i)}{\sum_i H(x_i) \cdot E_w(x_i)} \tag{A.19}$$

$$\sigma'_w = \frac{\sum_i (x_i - \mu'_w)^2 \cdot H(x_i) \cdot E_w(x_i)}{\sum_i H(x_i) \cdot E_w(x_i)} \tag{A.20}$$

$$a'_f = \frac{1}{N} \sum_i H(x_i) \cdot E_f(x_i) \tag{A.21}$$

$$\mu'_f = \frac{\sum_i x_i \cdot H(x_i) \cdot E_f(x_i)}{\sum_i H(x_i) \cdot E_f(x_i)} \tag{A.22}$$

$$\sigma'_f = \frac{\sum_i (x_i - \mu'_f)^2 \cdot H(x_i) \cdot E_f(x_i)}{\sum_i H(x_i) \cdot E_f(x_i)} \tag{A.23}$$

$$a'_b = \frac{1}{N} \sum_i H(x_i) \cdot E_b(x_i) \tag{A.24}$$

$$\mu'_b = \frac{\sum_i x_i \cdot H(x_i) \cdot E_b(x_i)}{\sum_i H(x_i) \cdot E_b(x_i)} \tag{A.25}$$

A.4 RF shimming with SAR constraints

Recall that the proposed cost function used for RF shimming with strict SAR constraints is defined as:

$$C(\mathbf{w}) = ||\mathbf{A}\mathbf{w} - \mathbf{b}||_2^2 - \frac{1}{t}\sum_i \ln(s - \mathbf{w}^H\mathbf{Q}_i\mathbf{w}) \qquad (A.26)$$

where $s := \mathrm{SAR}_{\max}/a_{\mathrm{rms}}^2$. As $C(\mathbf{w})$ is convex, the global minimum can be found iteratively, starting from any feasible initial solution (e.g. $\mathbf{w}_0 = \mathbf{0}$). For RF shimming, the number of unknowns is relatively small and Newton's method can be used to calculate the descent direction $\Delta\mathbf{w}_{n+1}$ at the n-th search step via the linear equation system:

$$\nabla^2 C(\mathbf{w}_n) \cdot \Delta\mathbf{w}_{n+1} = -\nabla C(\mathbf{w}_n) \qquad (A.27)$$

where the gradient $\nabla C(\mathbf{w})$ and the Hessian $\nabla^2 C(\mathbf{w})$ of the cost function are given by:

$$\nabla C(\mathbf{w}) = 2\mathbf{A}^H\mathbf{A}\mathbf{w} - 2\mathbf{A}^H\mathbf{b} + \frac{2}{t}\sum_i \frac{\mathbf{Q}_i\mathbf{w}}{s - \mathbf{w}^H\mathbf{Q}_i\mathbf{w}} \qquad (A.28)$$

$$\nabla^2 C(\mathbf{w}) = 2\mathbf{A}^H\mathbf{A} + \frac{2}{t}\sum_i \left[\frac{\mathbf{Q}_i}{s - \mathbf{w}^H\mathbf{Q}_i\mathbf{w}} + \frac{2\mathbf{Q}_i\mathbf{w}\mathbf{w}^H\mathbf{Q}_i^H}{(s - \mathbf{w}^H\mathbf{Q}_i\mathbf{w})^2} \right] $$
$$(A.29)$$

To calculate the updated solution $\mathbf{w}_{n+1}$, a feasible step length $l \in (0; 1]$ has to be found, such that the solution trajectory stays within the inner points of the barrier (e.g. via a simple line search). The updated solution can then be calculated as:

$$\mathbf{w}_{n+1} = \mathbf{w}_n + l \cdot \Delta\mathbf{w}_{n+1} \qquad (A.30)$$

For sequential optimization of the target phase in the magnitude least-squares (MLS) [110], the barrier method has to be applied for each iteration.

A.5 Tissue properties

This section lists the properties of 47 human body tissues. The dielectric properties (conductivity σ and relative permittivity ε_r) were calculated according to Gabriel *et al.* [84] at MR-relevant RF frequencies. Density values were assembled from [74, 86, 123, 124].

Tissue type	64 MHz		128 MHz		300 MHz		ρ in g/cm^3
	σ in S/m	ε_r	σ in S/m	ε_r	σ in S/m	ε_r	
Aorta	0.429	68.6	0.479	56.0	0.537	48.3	1.15
Bladder	0.287	24.6	0.298	21.9	0.317	20.1	1.11
Blood	1.207	86.4	1.249	73.2	1.316	65.7	1.06
Blood vessel	0.429	68.6	0.479	56.0	0.537	48.3	1.13
Bone (cancellous)	0.161	30.9	0.180	26.3	0.216	23.2	1.08
Bone (cortical)	0.060	16.7	0.067	14.7	0.083	13.4	1.85
Bone (marrow)	0.021	7.2	0.024	6.2	0.027	5.8	0.94
Brain (gray matter)	0.511	97.4	0.587	73.5	0.692	60.0	1.04
Brain (white matter)	0.292	67.8	0.342	52.5	0.413	43.8	1.03
Breast fat	0.030	5.8	0.030	5.6	0.033	5.5	1.15
Cartilage	0.452	62.9	0.488	52.9	0.553	46.8	1.10
Cerebellum	0.719	116.3	0.829	79.7	0.973	59.7	1.04
Cerebro-spinal fluid	2.066	97.3	2.143	84.0	2.224	72.7	1.00
Cervix	0.726	67.5	0.755	57.7	0.799	52.6	1.16
Colon	0.638	94.7	0.705	76.6	0.811	65.0	1.13
Cornea	1.001	87.4	1.059	71.5	1.151	61.4	1.08

Tissue type	64 MHz		128 MHz		300 MHz		ρ in g/cm^3
	σ in S/m	ε_r	σ in S/m	ε_r	σ in S/m	ε_r	
Eye sclera	0.883	75.3	0.918	65.0	0.975	58.9	1.15
Fat	0.035	6.5	0.037	5.9	0.040	5.6	0.96
Gall bladder	0.966	87.4	1.042	74.1	1.117	62.9	1.12
Gall bladder bile	1.482	105.4	1.576	88.9	1.670	74.9	0.93
Glands	0.778	74.0	0.804	66.8	0.851	62.5	1.05
Heart	0.678	106.5	0.766	84.3	0.904	69.3	1.06
Kidney	0.741	118.6	0.852	89.6	1.022	70.5	1.15
Lens	0.586	60.5	0.609	53.1	0.648	49.0	1.05
Liver	0.448	80.6	0.511	64.3	0.610	53.5	1.16
Lung (deflated)	0.531	75.3	0.576	63.7	0.649	56.2	0.56
Lung (inflated)	0.289	37.1	0.316	29.5	0.356	24.8	0.21
Lymph	0.778	74.0	0.804	66.8	0.851	62.5	1.04
Mucous membrane	0.488	76.7	0.544	61.6	0.631	51.9	1.04
Muscle	0.688	72.2	0.719	63.5	0.771	58.2	1.18
Nail	0.060	16.7	0.067	14.7	0.083	13.4	1.03
Nerve	0.312	55.1	0.354	44.1	0.418	36.9	1.11
Ovary	0.685	106.8	0.790	79.2	0.946	61.3	1.20
Pancreas	0.778	74.0	0.804	66.8	0.851	62.5	1.13
Prostate	0.885	84.5	0.926	72.1	0.994	64.8	1.14
Skin (dry)	0.436	92.2	0.523	65.4	0.641	49.8	1.13
Skin (wet)	0.488	76.7	0.544	61.6	0.631	51.9	1.13
Small intestine	1.591	118.4	1.693	88.0	1.841	69.8	1.15
Spleen	0.744	110.6	0.835	82.9	0.969	66.5	1.16

Tissue type	64 MHz		128 MHz		300 MHz		
	σ in S/m	ε_r	σ in S/m	ε_r	σ in S/m	ε_r	ρ in g/cm^3
Stomach	0.878	85.8	0.913	74.9	0.972	68.7	1.13
Tendon	0.474	59.5	0.499	51.9	0.537	48.0	1.20
Testis	0.885	84.5	0.926	72.1	0.994	64.8	1.12
Tongue	0.652	75.3	0.687	65.0	0.745	58.9	1.06
Tooth	0.060	16.7	0.067	14.7	0.083	13.4	2.16
Trachea	0.528	58.9	0.559	50.6	0.611	45.3	1.21
Uterus	0.911	92.1	0.961	75.4	1.038	66.2	1.16
Vitreous humor	1.503	69.1	1.505	69.1	1.518	69.0	1.00

Bibliography

[1] D. Hoult, "Sensitivity and power deposition in a high-field imaging experiment," *J. Magn. Reson. Imaging*, vol. 12, no. 1, pp. 46–67, 2000.

[2] T. Ibrahim, R. Lee, B. Baertlein, A. Abduljalil, H. Zhu, and P. Robitaille, "Effect of RF coil excitation on field inhomogeneity at ultra high fields: a field optimized TEM resonator," *Magn. Reson. Imaging*, vol. 19, no. 10, pp. 1339–1347, 2001.

[3] U. Katscher, P. Börnert, C. Leussler, and J. van den Brink, "Transmit SENSE," *Magn. Reson. Med.*, vol. 49, no. 1, pp. 144–150, 2003.

[4] Y. Zhu, "Parallel excitation with an array of transmit coils," *Magn. Reson. Med.*, vol. 51, no. 4, pp. 775–784, 2004.

[5] W. Grissom, C. Yip, Z. Zhang, V. Stenger, J. Fessler, and D. Noll, "Spatial domain method for the design of RF pulses in multicoil parallel excitation," *Magn. Reson. Med.*, vol. 56, no. 3, pp. 620–629, 2006.

[6] C. Collins and Z. Wang, "Caclulation of Radiofrequency Electromagnetic Fields and Their Effects in MRI of Human Subjects," *Magn. Reson. Med.*, vol. 65, no. 5, pp. 1470–1482, 2011.

[7] E. Haake, R. Brown, M. Thompson, and R. Venkatesan, *Magnetic Resonance Imaging: Physical Principles and Sequence Design*. Wiley, New York, 1999.

[8] C. Hayes, W. Edelstein, J. Schenck, O. Mueller, and M. Eash, "An efficient, highly homogeneous radiofrequency coil for whole-body NMR imaging at 1.5T," *J. Magn. Reson.*, vol. 63, pp. 622–628, 1985.

[9] J. Pauly, D. Nishimura, and A. Macovski, "A k-space analysis of small-tip-angle excitation," *J. Magn. Reson.*, vol. 81, no. 1, pp. 43–56, 1989.

[10] D. Twieg, "The k-trajectory formulation of the NMR imaging process with applications in analysis and synthesis of imaging methods," *Med. Phys.*, vol. 10, no. 5, pp. 610–621, 1983.

[11] V. Yarnykh, "Actual flip-angle imaging in the pulsed steady state: a method for rapid three-dimensional mapping of the transmitted radiofrequency field," *Magn. Reson. Med.*, vol. 57, no. 1, pp. 192–200, 2007.

[12] D. Hoult and R. Deslauriers, "Accurate shim-coil design and magnet-field profiling by a power-minimization-matrix method," *J. Magn. Reson., Ser. A*, vol. 108, no. 1, pp. 9–20, 1994.

[13] G. Yang, P. Gatehouse, J. Keegan, R. Mohiaddin, and D. Firmin, "Three-dimensional coronary MR angiography using zonal echo planar imaging," *Magn. Reson. Med.*, vol. 39, no. 5, pp. 833–842, 1998.

[14] T. Sachs, C. Meyer, B. Hu, J. Kohli, D. Nishimura, and A. Macovski, "Real-time motion detection in spiral MRI using navigators," *Magn. Reson. Med.*, vol. 32, no. 5, pp. 639–645, 1994.

[15] K. Prüssmann, M. Weiger, M. Scheidegger, and P. Bösiger, "SENSE: sensitivity encoding for fast MRI," *Magn. Reson. Med.*, vol. 42, no. 5, pp. 952–962, 1999.

[16] F. Shellock and E. Kanal, *Magnetic Resonance – Bioeffects, safety, and patient management*.   Raven Press, New York, 1994.

[17] International Electrotechnical Commission, "International standard, Medical equipment – IEC 60601-2-33: Particular requirements for the safety of Magnetic resonance equipment (3rd edition)," 2010.

[18] P. Bottomley, R. Redington, W. Edelstein, and J. Schenck, "Estimating radiofrequency power deposition in body NMR imaging," *Magn. Reson. Med.*, vol. 2, no. 4, pp. 336–349, 1985.

[19] C. Collins, L. Shizhe, and M. Smith, "SAR and B1 Field Distributions in a Heterogeneous Human Head Model within a Birdcage Coil," *Magn. Reson. Med.*, vol. 40, pp. 847–856, 1998.

[20] C. Gabriel, S. Gabriel, and E. Corthout, "The dielectric properties of biological tissues: I. Literature survey," *Phys. Med. Biol.*, vol. 41, pp. 2231–2249, 1996.

[21] M. Golombeck, "Feldtheoretische Studien zur Patientensicherheit bei der Magnetresonanztomographie und der Elektrochirurgie," Ph.D. dissertation, Institute of Biomedical Engineering (IBT), Karlsruhe Institute of Technology, 2003.

[22] H. Homann, I. Graesslin, H. Eggers, K. Nehrke, P. Vernickel, U. Katscher, O. Dössel, and P. Börnert, "Local SAR Management by RF Shimming: A Simulation Study with Multiple Human Body Models," *Magn. Reson. Mater. Phy.*, vol. accepted, 2011.

[23] H. Homann, I. Graesslin, T. Voigt, P. Börnert, and O. Dössel, "The influence of body size on the specific absorption rate (SAR)," in *Proceedings of the ESMRMB, Antalya, Turkey*, 2009, p. 522, (Magma 22, Suppl 1).

[24] International Commission on Non-Ionizing Radiation (ICNIRP), "Guidelines for limiting exposure to time-varying electric, magnetic, and electromagnetic fields (up to 300 GHz)," *Health Phys.*, vol. 74, no. 4, pp. 494–522, 1998.

[25] IEEE, "IEEE Standard for Safety Levels with Respect to Human Exposure to Radio Frequency Electromagnetic Fields, 3 kHz to 300 GHz," IEEE Std C95.3-2002, 2008.

[26] D. Schaefer, "IEC MR Safety Standard Development," in *ISMRM Workshop on RF safety, Stillwater*, 2010.

[27] C. Durney, C. Johnson, and P. Barber, *Radiofrequency radiation dosimetry handbook.* USAF School of aerospace medicine, Report SAM TR 78 22, Brooks Air Force Base, Texas, 1978.

[28] A. Boss, H. Graf, A. Berger, U. Lauer, H. Wojtczyk, C. Claussen, and F. Schick, "Tissue warming and regulatory responses induced

by radio frequency energy deposition on a whole-body 3-Tesla magnetic resonance imager," *J. Magn. Reson. Imaging*, vol. 26, no. 5, pp. 1334–1339, 2007.

[29] C. Collins, W. Liu, J. Wang, R. Gruetter, J. Vaughan, K. Ugurbil, and M. Smith, "Temperature and SAR calculations for a human head within volume and surface coils at 64 and 300 MHz," *J. Magn. Reson. Imaging*, vol. 19, no. 5, pp. 650–656, 2004.

[30] J. De Wilde, "Special Patient Groups e.g. Pregnancy, Diabetes and exposure to RF heating during MRI," in *ISMRM Workshop on RF safety, Lisbon*, 2008.

[31] National Electrical Manufacturers Association, "NEMA Standards Publication MS-8, Characterization of the Specific Absorption Rate for Magnetic Resonance Imaging Systems," 2008.

[32] L. Alon, C. Deniz, R. Lattanzi, G. Wiggins, R. Brown, D. Sodickson, and Y. Zhu, "An Automated Method for Subject Specific Global SAR Prediction in Parallel Transmission," in *Proceedings of the ISMRM, Stockholm, Sweden*, 2010, p. 780.

[33] M. Konings, L. Bartels, H. Smits, and C. Bakker, "Heating around intravascular guidewires by resonating RF waves," *J. Magn. Reson. Imaging*, vol. 12, no. 1, pp. 79–85, 2000.

[34] P. Hardy and K. Weil, "A Review of Thermal MR Injuries," *Radiol. Technol.*, vol. 81, no. 6, p. 606, 2010.

[35] U. Katscher, T. Voigt, C. Findeklee, P. Vernickel, K. Nehrke, and O. Dössel, "Determination of Electric Conductivity and Local SAR via B1 Mapping," *IEEE Trans. Med. Imaging*, vol. 28, no. 9, pp. 1365–1374, 2009.

[36] C. Collins, W. Mao, W. Liu, and M. Smith, "Calculated Local and Average SAR in Comparison with Regulatory Limits," in *Proceedings of the ISMRM, Seattle, WA, USA*, 2006, p. 2044.

[37] Z. Wang, J. Lin, W. Mao, W. Liu, M. Smith, and C. Collins, "SAR and temperature: simulations and comparison to regulatory limits for MRI," *J. Magn. Reson. Imaging*, vol. 26, no. 2, pp. 437–441, 2007.

[38] X. Chen, Y. Hamamura, and M. Steckner, "Numerical simulation of SAR for 3T whole body coil: Effect of patient loading positions on local SAR hotspot," in *Proceedings of the ISMRM, Stockholm, Sweden*, 2010, p. 3879.

[39] D. Yeo, Z. Wang, W. Loew, M. Vogel, and I. Hancu, "Local specific absorption rate in high-pass birdcage and transverse electromagnetic body coils for multiple human body models in clinical landmark positions at 3T," *J. Magn. Reson. Imaging*, vol. 33, no. 5, pp. 1209–1217, 2011.

[40] J. Nadobny, M. Szimtenings, D. Diehl, E. Stetter, G. Brinker, and P. Wust, "Evaluation of MR-induced hot spots for different temporal SAR modes using a time-dependent finite difference method with explicit temperature gradient treatment," *IEEE Trans. Biomed. Eng.*, vol. 54, no. 10, pp. 1837–1850, 2007.

[41] Z. Wang, J. Lin, J. Vaughan, and C. Collins, "Consideration of Physiological Response in Numerical Models of Temperature during MRI of the Human Head," *J. Magn. Reson. Imaging*, vol. 28, pp. 1303–1308, 2008.

[42] A. van Lier, C. van den Berg, D. Klomp, B. Raaymakers, and J. Lagendijk, "RF safety assessment of a 7T head coil using thermal modeling with discrete vessels," in *Proceedings of the ISMRM, Honolulu, Hawaii, USA*, 2009, p. 303.

[43] V. Rieke and K. Butts Pauly, "MR thermometry," *J. Magn. Reson. Imaging*, vol. 27, no. 2, pp. 376–390, 2008.

[44] B. de Senneville, C. Mougenot, B. Quesson, I. Dragonu, N. Grenier, and C. Moonen, "MR thermometry for monitoring tumor ablation," *European radiology*, vol. 17, no. 9, pp. 2401–2410, 2007.

[45] S. Oh, Y. Ryu, A. Webb, and C. Collins, "In-vivo Human Forearm Temperature Mapping for Correspondence with Numerical SAR and Temperature Calculations," in *Proceedings of the ISMRM, Montreal, Canada*, 2011, p. 3863.

[46] F. Seifert, G. Wubbeler, S. Junge, B. Ittermann, and H. Rinneberg, "Patient safety concept for multichannel transmit coils," *J. Magn. Reson. Imaging*, vol. 26, no. 5, pp. 1315–21, 2007.

[47] C. Collins, Z. Wang, and M. Smith, "A conservative method for ensuring safety within transmit arrays," in *Proceedings of the ISMRM, Berlin, Germany*, 2007, p. 1092.

[48] I. Graesslin, D. Glaesel, S. Biederer, P. Vernickel, U. Katscher, F. Schweser, B. Annighoefer, H. Dingemans, G. Mens, G. van Yperen, and P. Harvey, "Comprehensive RF Safety Concept for Parallel Transmission Systems," in *Proceedings of the ISMRM, Toronto, Canada*, 2008, p. 74.

[49] D. Brunner, J. Paska, J. Fröhlich, and K. Prüssmann, "SAR assessment of transmit arrays: Deterministic calculation of worst- and best-case performance," in *Proceedings of the ISMRM, Honolulu, Hawaii, USA*, 2009, p. 4803.

[50] F. Bardati, A. Borrani, A. Gerardino, and G. Lovisolo, "SAR optimization in a phased array radiofrequency hyperthermia system," *IEEE Trans. Biomed. Eng.*, vol. 42, no. 12, pp. 1201–1207, 1995.

[51] M. Gebhardt, D. Diehl, E. Adalsteinsson, L. Wald, and G. Eichfelder, "Evaluation of maximum local SAR for parallel transmission (pTx) pulses based on pre-calculated field data using a selected subset of Virtual Observation Points," in *Proceedings of the ISMRM, Stockholm, Sweden*, 2010, p. 1441.

[52] G. Eichfelder and M. Gebhardt, "Local Specific Absorption Rate Control for Parallel Transmission by Virtual Observation Points," *Magn. Reson. Med.*, vol. 66, no. 5, pp. 1468–1476, 2011.

[53] P. Vernickel, P. Röschmann, C. Findeklee, K. Lüdeke, C. Leussler, J. Overweg, U. Katscher, I. Grasslin, and K. Schünemann, "Eight-channel transmit/receive body MRI coil at 3T," *Magn. Reson. Med.*, vol. 58, no. 2, pp. 381–389, 2007.

[54] I. Graesslin, P. Vernickel, J. Schmidt, C. Findeklee, P. Roeschmann, C. Leussler, P. Haaker, H. Laudan, K. Luedeke, J. Scholz, S. Buller, J. Keupp, P. Börnert, H. Dingemans, G. Mens, K. Blom, N. Swennen, L. Mollevanger, P. Harvey, and K. U., "Whole Body 3T MRI System with Eight Parallel RF Transmission Channels," in *Proceedings of the ISMRM, Seattle, WA, USA*, 2006, p. 129.

[55] A. Kost, *Numerische Methoden in der Berechnung elektromagnetischer Felder.* Springer, 1994.

[56] K. Yee, "Numerical solution of initial boundary value problems involving Maxwell's equations in isotropic media," *IEEE Trans. Antennas Propag.*, vol. 14, no. 3, pp. 302–307, 1966.

[57] K. Kunz and R. Luebbers, *The Finite Difference Time Domain Method for Electromagnetics.* CRC press, Boca Raton, Florida, 1993.

[58] J. Vaughan, H. Hetherington, J. Otu, J. Pan, and G. Pohost, "High frequency volume coils for clinical NMR imaging and spectroscopy," *Magn. Reson. Med.*, vol. 32, no. 2, pp. 206–218, 1994.

[59] K. Falaggis, "Computation and Validation of Electromagnetic fields and Radiofrequency Exposure for Parallel Transmission," Master's thesis, Hamburg University of Technology, 2006.

[60] C. Penney, "Computation of Fields and SAR for MRI with Finite-Difference Time-Domain Software," *Mircowave Journal*, vol. 50, no. 12, pp. 118–122, 2007.

[61] G. McKinnon and Z. Wang, "Direct Capacitor Determination in FDTD Modeling of RF Coils," in *Proceedings of the ISMRM, Toronto, Canada*, 2003, p. 2381.

[62] M. Kozlov and R. Turner, "Fast MRI coil analysis based on 3-D electromagnetic and RF circuit co-simulation," *J. Magn. Reson.*, vol. 200, no. 1, pp. 147–152, 2009.

[63] I. Graesslin, S. Biederer, K. Falaggis, H. Vernickel, P. nd Dingemans, G. Mens, P. Röschmann, C. Leussler, Z. Zhai, M. Morich, and U. Katscher, "Real-time SAR Monitoring to ensure Patient Safety for Parallel Transmission Systems," in *Proceedings of the ISMRM, Berlin, Germany*, 2007, p. 1086.

[64] P. Vernickel, C. Findeklee, J. Eichmann, and I. Graesslin, "Active digital decoupling for multi-channel transmit MRI Systems," in *Proceedings of the ISMRM, Berlin, Germany*, 2007, p. 170.

[65] G. Scott, P. Stang, W. Overall, A. Kerr, and J. Pauly, "Signal Vector Decoupling for Transmit Arrays," in *Proceedings of the ISMRM, Berlin, Germany*, 2007, p. 168.

[66] V. Yarnykh and C. Yuan, "Actual Flip Angle Imaging in the Pulsed Steady State," in *Proceedings of the ISMRM, Kyoto, Japan*, vol. 11, 2004, p. 194.

[67] K. Nehrke, "On the steady-state properties of actual flip angle imaging (AFI)," *Magn. Reson. Med.*, vol. 61, no. 1, pp. 84–92, 2009.

[68] D. Brunner and K. Prüssmann, "B1+ interferometry for the calibration of RF transmitter arrays," *Magn. Reson. Med.*, vol. 61, no. 6, pp. 1480–1488, 2009.

[69] K. Nehrke and P. Börnert, "Eigenmode analysis of transmit coil array for tailored B1 mapping," *Magn. Reson. Med.*, vol. 63, pp. 754–764, 2010.

[70] H. Taylor, M. Burl, and J. Hand, "Design and calibration of electric field probes in the range 10-120 MHz," *Phys. Med. Biol.*, vol. 42, pp. 1387–1394, 1997.

[71] A. Stogryn, "Equations for calculating the dielectric constant of saline water," *IEEE Trans. Microwave Theory Tech.*, vol. 19, no. 8, pp. 733–736, 1971.

[72] National Library of Medicine (NLM), National Institutes of Health (NIH), Bethesda, MD, USA, "Visible Human Project," 1996.

[73] C. Collins and M. Smith, "Calculations of B1 distribution, SNR, and SAR for a surface coil adjacent to an anatomically-accurate human body model," *Magn. Reson. Med.*, vol. 45, no. 4, pp. 692–699, 2001.

[74] W. Liu, C. Collins, and M. Smith, "Calculations of B1, Distribution, Specific Energy Absorption Rate, and Intrinsic Signal-to-Noise Ratio for a Body-Size Birdcage Coil Loaded with Different Human Subjects at 64 and 128MHz," *Appl. Magn. Reson.*, vol. 29, no. 1, pp. 5–18, 2005.

[75] A. Christ, W. Kainz, E. Hahn, K. Honegger, M. Zefferer, E. Neufeld, W. Rascher, R. Janka, W. Bautz, J. Chen, B. Kiefer, P. Schmitt, H. Hollenbach, J. Shen, M. Oberle, D. Szczerba, A. Kam, J. Guag, and N. Kuster, "The Virtual Family – development of surface-based anatomical models of two adults and two children for dosimetric simulations," *Phys. Med. Biol.*, vol. 55, pp. N23–N38, 2010.

[76] P. Dimbylow, "FDTD calculations of the whole-body averaged SAR in an anatomically realistic voxel model of the human body from 1 MHz to 1 GHz," *Phys. Med. Biol.*, vol. 42, pp. 479–490, 1997.

[77] P. Dimbylow, "Development of the female voxel phantom, NAOMI, and its application to calculations of induced current densities and electric fields from applied low frequency magnetic and electric fields," *Phys. Med. Biol.*, vol. 50, pp. 1047–1070, 2005.

[78] Z. Zhai, M. Morich, G. DeMeester, and P. Harvey, "A Study of the Relationship between B1-field Uniformity, Body Aspect Ratio and SAR for Whole-Body RF Shimming at 3.0T," in *Proceedings of the ISMRM, Honolulu, Hawaii, USA*, 2009, p. 3045.

[79] B. van den Bergen, C. van den Berg, H. Kroeze, L. Bartels, and J. Lagendijk, "The effect of body size and shape on RF safety and B1 field homogeneity at 3T," in *Proceedings of the ISMRM, Seattle, WA, USA*, 2006, p. 2040.

[80] Z. Wang, D. Yeo, C. Collins, J. Jin, and F. Robb, "SAR Comparison for Multiple Human Body Models at 1.5T and 3.0T," in *Proceedings of the ISMRM, Stockholm, Sweden*, 2010, p. 3880.

[81] Z. Wang, C. Penney, R. Luebbers, and C. Collins, "Poseable Male and Female Numerical Body Models for Field Calculations in MRI," in *Proceedings of the ISMRM, Toronto, Canada*, 2008, p. 75.

[82] Z. Wang, C. Collins, S. Zhao, and F. Robb, "The Effect of Human Model Resolution on Numerical Calculation of SAR and Temperature in MRI," in *Proceedings of the ISMRM, Honolulu, Hawaii, USA*, 2009, p. 4797.

[83] C. Collins and M. Smith, "Spatial Resolution of Numerical Models of Man and Calculated Specific Absorption Rate Using the FDTD Method: A Study at 64MHz in a Magnetic Resonance Imaging Coil," *J. Magn. Reson. Imaging*, vol. 18, no. 3, pp. 383–388, 2003.

[84] S. Gabriel, R. Lau, and C. Gabriel, "The dielectric properties of biological tissues: III. Parametric models for the dielectric spectrum of tissues," *Phys. Med. Biol.*, vol. 41, pp. 2271–2293, 1996.

[85] H. Homann, P. Börnert, H. Eggers, K. Nehrke, O. Dössel, and I. Graesslin, "Toward Individualized SAR Models and In Vivo Validation," *Magn. Reson. Med.*, vol. 66, pp. 1767–1776, 2011.

[86] P. Mason, W. Hurt, T. Walters, J. D'Andrea, P. Gajsek, K. Ryan, D. Nelson, K. Smith, and J. Ziriax, "Effects of frequency, permittivity, and voxel size on predicted specific absorption rate values in biological tissue during electromagnetic-field exposure," *IEEE Trans. Microwave Theory Tech.*, vol. 48, no. 11 Part 2, pp. 2050–2058, 2000.

[87] C. Gabriel and A. Peyman, "Dielectric measurement: error analysis and assessment of uncertainty," *Phys. Med. Biol.*, vol. 51, no. 23, pp. 6033–6046, 2006.

[88] E. Conil, A. Hadjem, F. Lacroux, M. Wong, and J. Wiart, "Variability analysis of SAR from 20 MHz to 2.4 GHz for different adult and child models using finite-difference time-domain," *Phys. Med. Biol.*, vol. 53, no. 6, pp. 1511–1526, 2008.

[89] B. van den Bergen, C. van den Berg, L. Bartels, and J. Lagendijk, "7T body MRI: B1 shimming with simultaneous SAR reduction," *Phys. Med. Biol.*, vol. 52, no. 17, pp. 5429–5442, 2007.

[90] P. Gajsek, W. Hurt, J. Ziriax, and P. Mason, "Parametric dependence of SAR on permittivity values in a man model," *IEEE Trans. Biomed. Eng.*, vol. 48, no. 10, pp. 1169–1177, 2001.

[91] J. Ma, "Dixon techniques for water and fat imaging," *J. Magn. Reson. Imaging*, vol. 28, no. 3, pp. 543–558, 2008.

[92] P. Koken, H. Eggers, and P. Börnert, "Fast Single Breath-Hold 3D Abdominal Imaging with Water-Fat Separation," in *Proceedings of the ISMRM, Berlin, Germany*, 2007, p. 1623.

[93] S. Reeder, Z. Wen, H. Yu, A. Pineda, G. Gold, M. Markl, and N. Pelc, "Multicoil Dixon chemical species separation with an iterative least-squares estimation method," *Magn. Reson. Med.*, vol. 51, no. 1, pp. 35–45, 2004.

[94] H. Eggers, B. Brendel, A. Duijndam, and G. Herigault, "Dual-Echo Dixon Imaging with Flexible Choice of Echo Times," *Magn. Reson. Med.*, vol. 65, pp. 96–107, 2011.

[95] D. Wilson and J. Noble, "Segmentation of cerebral vessels and aneurysms from MR angiography data," in *Proc. Information Processing in Medical Imaging (IPMI)*, 1997, pp. 423–428.

[96] P. Börnert, J. Keupp, H. Eggers, and B. Aldefeld, "Whole-body 3D water/fat resolved continuously moving table imaging," *J. Magn. Reson. Imaging*, vol. 25, no. 3, pp. 660–665, 2007.

[97] M. Doneva, P. Börnert, H. Eggers, A. Mertins, J. Pauly, and M. Lustig, "Compressed Sensing for Chemical Shift based Water-Fat Separation," *Magn. Reson. Med.*, vol. 64, pp. 1114–1120, 2010.

[98] C. van den Berg, L. Bartels, B. van den Bergen, H. Kroeze, A. de Leeuw, J. Van de Kamer, and J. Lagendijk, "The use of MR B1+ imaging for validation of FDTD electromagnetic simulations of human anatomies," *Phys. Med. Biol.*, vol. 51, pp. 4735–4746, 2006.

[99] M. Hofmann, B. Pichler, B. Schölkopf, and T. Beyer, "Towards quantitative PET/MRI: A review of MR-based attenuation correction techniques," *Eur. J. Nucl. Med. Mol. Imaging*, vol. 36, pp. 93–104, 2009.

[100] S. Wang and J. Duyn, "Fast SAR estimation via a Hybrid Approach," in *Proceedings of the ISMRM, Stockholm, Sweden*, 2010, p. 1448.

[101] C. van den Berg, B. van den Bergen, J. van de Kamer, B. Raaymakers, H. Kroeze, L. Bartels, and J. Lagendijk, "Simultaneous B1+ homogenization and specific absorption rate hotspot suppression using a magnetic resonance phased array transmit coil," *Magn. Reson. Med.*, vol. 57, no. 3, pp. 577–586, 2007.

[102] P. Harvey, Z. Zhai, M. Morich, G. Mens, G. van Yperen, G. De-Meester, I. Graesslin, and R. Hoogeveen, "SAR Behavior During Whole-Body MultiTransmit RF Shimming at 3.0T," in *Proceedings of the ISMRM, Honolulu, Hawaii, USA*, 2009, p. 4786.

[103] S. Buchenau, M. Haas, J. Hennig, and M. Zaitsev, "A comparison of local SAR using individual patient data and a patient template," in *Proceedings of the ISMRM, Honolulu, Hawaii, USA*, 2009, p. 4798.

[104] M. Cloos, M. Luong, G. Ferrand, A. Amadon, D. Le Bihan, and N. Boulant, "Local SAR reduction in parallel excitation based on channel-dependent Tikhonov parameters," *J. Magn. Reson. Imaging*, vol. 32, no. 5, pp. 1209–1216, 2010.

[105] J. Lee, M. Gebhardt, L. Wald, and E. Adalsteinsson, "Parallel Transmit RF Design with Local SAR Constraints," in *Proceedings of the ISMRM, Stockholm, Sweden*, 2010, p. 105.

[106] A. Sbrizzi, H. Hoogduin, J. Lagendijk, P. Luijten, G. Sleijpen, and C. van den Berg, "A Fast Algorithm for local-1gram-SAR optimized Parallel-Transmit RF-Pulse Design," in *Proceedings of the ISMRM, Stockholm, Sweden*, 2010, p. 4931.

[107] A. Zelinski, V. Goyal, L. Angelone, G. Bonmassar, L. Wald, and E. Adalsteinsson, "Designing RF pulses with optimal specific absorption rate (SAR) characteristics and exploring excitation fidelity, SAR and pulse duration tradeoffs," in *Proceedings of the ISMRM, Berlin, Germany*, 2007, p. 1699.

[108] D. Brunner and K. Prüssmann, "Optimal design of multiple-channel RF pulses under strict power and SAR constraints," *Magn. Reson. Med.*, vol. 63, no. 5, pp. 1280–1291, 2010.

[109] S. Boyd and L. Vandenberghe, *Convex Optimization*. Cambridge University Press, 2009.

[110] K. Setsompop, L. Wald, V. Alagappan, B. Gagoski, and E. Adalsteinsson, "Magnitude least squares optimization for parallel radio frequency excitation design demonstrated at 7 Tesla with eight channels," *Magn. Reson. Med.*, vol. 59, no. 4, pp. 908–915, 2008.

[111] P. Kassakian, "Convex approximation and optimization with applications in magnitude filter design and radiation pattern synthesis," Ph.D. dissertation, University of California at Berkeley, 2006.

[112] Z. Wang, S. Oh, M. Smith, and C. Collins, "RF Shimming Considering Both Excitation Homogeneity and SAR," in *Proceedings of the ISMRM, Berlin, Germany*, 2007, p. 1022.

[113] M. Fuderer, "The Information Content of MR Images," *IEEE Trans. Med. Imaging*, vol. 7, pp. 368–380, 1988.

[114] J. Van Vaals, M. Brummer, W. Thomas Dixon, H. Tuithof, H. Engels, R. Nelson, B. Gerety, J. Chezmar, and J. Den Boer, "Keyhole method for accelerating imaging of contrast agent uptake," *J. Magn. Reson. Imaging*, vol. 3, no. 4, pp. 671–675, 1993.

[115] R. Jones, O. Haraldseth, T. Müller, P. Rinck, and A. Øksendal, "K-space substitution: A novel dynamic imaging technique," *Magn. Reson. Med.*, vol. 29, no. 6, pp. 830–834, 1993.

[116] T. Schäffter, S. Weiss, and P. Börnert, "A SAR-reduced steady state free precessing (SSFP) acquisition," in *Proceedings of the ISMRM, Honolulu, Hawaii, USA*, 2002, p. 2351.

[117] J. Hennig and K. Scheffler, "Hyperechoes," *Magn. Reson. Med.*, vol. 46, no. 1, pp. 6–12, 2001.

[118] C. Lin, M. Bernstein, G. Gibbs, and J. Huston, "Reduction of RF power for magnetization transfer with optimized application of RF pulses in k-space," *Magn. Reson. Med.*, vol. 50, no. 1, pp. 114–121, 2003.

[119] P. Börnert, J. Weller, and I. Graesslin, "SAR Reduction in Parallel Transmission by k-space Dependent RF Pulse Selection," in *Proceedings of the ISMRM, Honolulu, Hawaii, USA*, 2009, p. 2600.

[120] U. Katscher, P. Vernickel, I. Graesslin, and P. Börnert, "RF shimming using a multi-element transmit system in phantom and in vivo studies," in *Proceedings of the ISMRM, Berlin, Germany*, 2007, p. 1693.

[121] D. Xu and K. King, "Joint Design of Excitation and Refocusing Pulses for Fast Spin Echo Sequences in Parallel Transmission," in *Proceedings of the ISMRM, Honolulu, Hawaii, USA*, 2009, p. 174.

[122] D. Hoult, "The Principle of Reciprocity in Signal Strength Calculations – A Mathematical Guide," *Concepts Magn. Reson.*, vol. 12, no. 4, pp. 173–187, 2000.

[123] W. Erdmann and T. Gos, "Density of trunk tissues of young and medium age people," *J. Biomech.*, vol. 23, no. 9, pp. 945–947, 1990.

[124] J. van Dyk, T. Keane, and W. Rider, "Lung density as measured by computerized tomography: implications for radiotherapy," *Int. J. Radiat. Oncol. Biol. Phys.*, vol. 8, no. 8, pp. 1363–1372, 1982.

List of publications

Peer-reviewed papers

- **H. Homann**, I. Graesslin, H. Eggers, K. Nehrke, P. Vernickel, U. Katscher, O. Dössel, and P. Börnert, "Local SAR Management by RF shimming: A simulation study with multiple human body models," *Magn. Reson. Mater. Phy.*, accepted, 2011. *Young Investigator Award 2011 of the ESMRMB (2nd price).*

- **H. Homann**, P. Börnert, H. Eggers, K. Nehrke, O. Dössel, and I. Graesslin, "Towards Individualized SAR Models and in vivo Validation," *Magn. Reson. Med.*, vol. 66, pp. 1767–1776, 2011.

- **H. Homann**, I. Graesslin, K. Nehrke, C. Findeklee, O. Dössel, and P. Börnert, "Specific Absorption Rate Reduction by k-Space Adaptive Radiofrequency Pulse Design," *Magn. Reson. Med.*, vol. 65, pp. 350–357, 2011.

Conference contributions

- **H. Homann**, I. Graesslin, H. Eggers, K. Nehrke, P. Vernickel, U. Katscher, O. Dössel, and P. Börnert, "Local SAR Management by RF shimming: A simulation study with multiple human body models," in *Proceedings of the ESMRMB, Leipzig, Germany*, 2011.

- **H. Homann**, P. Börnert, K. Nehrke, H. Eggers, O. Dössel, and I. Graesslin, "Validation and Comparison of Patient-Specific SAR Models," in *Proceedings of the ISMRM, Montreal*, Canada, p. 489, 2011.

- **H. Homann**, P. Börnert, O. Dössel, and I. Graesslin, "A Robust Concept for Real-Time SAR Calculation in Parallel Transmission," in *Proceedings of the ISMRM, Montreal, Canada*, p. 3843, 2011.

- **H. Homann**, T. Nielsen, K. Nehrke, I. Graesslin, O. Dössel, and P. Börnert, "Optimized RX Field Homogeneity for SENSE Imaging in Parallel Transmit MR," in *Proceedings of the ISMRM, Montreal, Canada*, p. 4404, 2011.

- **H. Homann**, I. Graesslin, H. Eggers, K. Nehrke, P. Börnert, and O. Dössel, "Patient-specific SAR models and in vivo validation," in *Proceedings of the ISMRM, Stockholm, Sweden*, p. 3874, 2010.

- **H. Homann**, K. Nehrke, I. Graesslin, O. Dössel, and P. Börnert, "SAR reduction by k-space adaptive RF shimming," in *Proceedings of the ISMRM, Stockholm, Sweden*, p. 104, 2010.

- **H. Homann**, I. Graesslin, T. Voigt, P. Börnert, and O. Dössel, "Towards patient-specific SAR models: A study on spatial and dielectric model resolution," in *Proceedings of the ESMRMB, Antalya, Turkey*, p. 195, 2009.

- **H. Homann**, I. Graesslin, T. Voigt, P. Börnert, and O. Dössel, "The influence of body size on the specific absorption rate (SAR)," in *Proceedings of the ESMRMB, Antalya, Turkey*, p. 522, 2009.

Karlsruhe Transactions on Biomedical Engineering
(ISSN 1864-5933)

Karlsruhe Institute of Technology / Institute of Biomedical Engineering (Ed.)

Die Bände sind unter www.ksp.kit.edu als PDF frei verfügbar oder als Druckausgabe bestellbar.

Band 2   Matthias Reumann
Computer assisted optimisation on non-pharmacological treatment of congestive heart failure and supraventricular arrhythmia. 2007
ISBN 978-3-86644-122-4

Band 3   Antoun Khawaja
Automatic ECG analysis using principal component analysis and wavelet transformation. 2007
ISBN 978-3-86644-132-3

Band 4   Dmytro Farina
Forward and inverse problems of electrocardiography : clinical investigations. 2008
ISBN 978-3-86644-219-1

Band 5   Jörn Thiele
Optische und mechanische Messungen von elektrophysiologischen Vorgängen im Myokardgewebe. 2008
ISBN 978-3-86644-240-5

Band 6   Raz Miri
Computer assisted optimization of cardiac resynchronization therapy. 2009
ISBN 978-3-86644-360-0

Band 7   Frank Kreuder
2D-3D-Registrierung mit Parameterentkopplung für die Patientenlagerung in der Strahlentherapie. 2009
ISBN 978-3-86644-376-1

Band 8   Daniel Unholtz
Optische Oberflächensignalmessung mit Mikrolinsen-Detektoren für die Kleintierbildgebung. 2009
ISBN 978-3-86644-423-2

Band 9   Yuan Jiang
Solving the inverse problem of electrocardiography in a realistic environment. 2010
ISBN 978-3-86644-486-7

Karlsruhe Transactions on Biomedical Engineering
(ISSN 1864-5933)

Band 10   Sebastian Seitz
Magnetic Resonance Imaging on Patients with Implanted Cardiac Pacemakers. 2011
ISBN 978-3-86644-610-6

Band 11   Tobias Voigt
Quantitative MR Imaging of the Electric Properties and Local SAR based on Improved RF Transmit Field Mapping. 2011
ISBN 978-3-86644-598-7

Band 12   Frank Michael Weber
Personalizing Simulations of the Human Atria: Intracardiac Measurements, Tissue Conductivities, and Cellular Electrophysiology. 2011
ISBN 978-3-86644-646-5

Band 13   David Urs Josef Keller
Multiscale Modeling of the Ventricles: from Cellular Electrophysiology to Body Surface Electrocardiograms. 2011
ISBN 978-3-86644-714-1

Band 14   Oussama Jarrousse
Modified Mass-Spring System for Physically Based Deformation Modeling. 2012
ISBN 978-3-86644-742-4

Band 15   Julia Bohnert
Effects of Time-Varying Magnetic Fields in the Frequency Range 1 kHz to 100 kHz upon the Human Body: Numerical Studies and Stimulation Experiment. 2012
ISBN 978-3-86644-782-0

Band 16   Hanno Homann
SAR Prediction and SAR Management for Parallel Transmit MRI. 2012
ISBN 978-3-86644-800-1